This report contains the collective views of an internation[illegible] does not necessarily represent the decisions or the stated p[illegible] of the United Nations Environment Programme, the International Labour Organization or the World Health Organization.

Environmental Health Criteria 233

TRANSGENIC ANIMAL MUTAGENICITY ASSAYS

First draft prepared by Drs Ulrich Wahnschaffe, Janet Kielhorn, Annette Bitsch and Inge Mangelsdorf, Fraunhofer Institute of Toxicology and Experimental Medicine, Hanover, Germany

Published under the joint sponsorship of the United Nations Environment Programme, the International Labour Organization and the World Health Organization, and produced within the framework of the Inter-Organization Programme for the Sound Management of Chemicals.

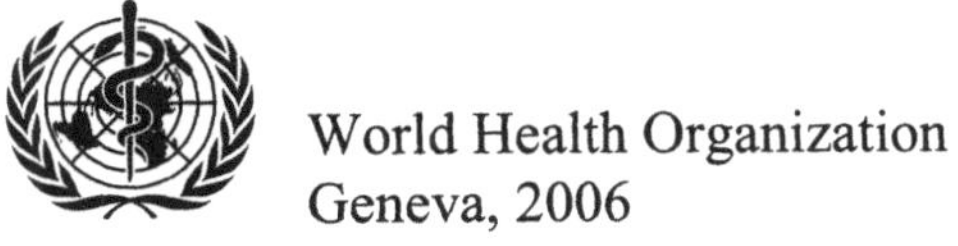

World Health Organization
Geneva, 2006

The **International Programme on Chemical Safety (IPCS)**, established in 1980, is a joint venture of the United Nations Environment Programme (UNEP), the International Labour Organization (ILO) and the World Health Organization (WHO). The overall objectives of the IPCS are to establish the scientific basis for assessment of the risk to human health and the environment from exposure to chemicals, through international peer review processes, as a prerequisite for the promotion of chemical safety, and to provide technical assistance in strengthening national capacities for the sound management of chemicals.

The **Inter-Organization Programme for the Sound Management of Chemicals (IOMC)** was established in 1995 by UNEP, ILO, the Food and Agriculture Organization of the United Nations, WHO, the United Nations Industrial Development Organization, the United Nations Institute for Training and Research and the Organisation for Economic Co-operation and Development (Participating Organizations), following recommendations made by the 1992 UN Conference on Environment and Development to strengthen cooperation and increase coordination in the field of chemical safety. The purpose of the IOMC is to promote coordination of the policies and activities pursued by the Participating Organizations, jointly or separately, to achieve the sound management of chemicals in relation to human health and the environment.

WHO Library Cataloguing-in-Publication Data

Transgenic animal mutagenicity assays.

(Environmental health criteria ; 233)

1.Mutagenicity tests - methods. 2.Mice, Transgenic. 3.Risk assessment. 4.Environmental exposure. I.World Health Organization. II.International Programme on Chemical Safety.

ISBN 92 4 157233 7 (NLM classification: QU 450)
ISBN 978 92 4 157233 7
ISSN 0250-863X

©World Health Organization 2006

All rights reserved. Publications of the World Health Organization can be obtained from WHO Press, World Health Organization, 20 Avenue Appia, 1211 Geneva 27, Switzerland (tel: +41 22 791 3264; fax: +41 22 791 4857; email: bookorders@who.int). Requests for permission to reproduce or translate WHO publications — whether for sale or for noncommercial distribution — should be addressed to WHO Press, at the above address (fax: +41 22 791 4806; email: permissions@who.int).

The designations employed and the presentation of the material in this publication do not imply the expression of any opinion whatsoever on the part of the World Health Organization concerning the legal status of any country, territory, city or area or of its authorities, or concerning the delimitation of its frontiers or boundaries. Dotted lines on maps represent approximate border lines for which there may not yet be full agreement.

The mention of specific companies or of certain manufacturers' products does not imply that they are endorsed or recommended by the World Health Organization in preference to others of a similar nature that are not mentioned. Errors and omissions excepted, the names of proprietary products are distinguished by initial capital letters.

All reasonable precautions have been taken by WHO to verify the information contained in this publication. However, the published material is being distributed without warranty of any kind, either express or implied. The responsibility for the interpretation and use of the material lies with the reader. In no event shall the World Health Organization be liable for damages arising from its use.

This document was technically and linguistically edited by Marla Sheffer, Ottawa, Canada, and printed by Wissenchaftliche Verlagsgesellschaft mbH, Stuttgart, Germany.

CONTENTS

ENVIRONMENTAL HEALTH CRITERIA FOR TRANSGENIC ANIMAL MUTAGENICITY ASSAYS

PREAMBLE x

ACRONYMS AND ABBREVIATIONS xix

GLOSSARY xxii

1. SUMMARY 1

PART I: OVERVIEW OF GENOTOXICITY TESTING AND TRANSGENIC ANIMAL MUTAGENICITY SYSTEMS 7

2. OVERVIEW OF GENOTOXICITY TESTING 8

   2.1 Gene mutation assays in vivo using endogenous genes 8
   2.2 Gene mutation assays in vivo using transgenes 8

3. CONSTRUCTION/PRODUCTION OF TRANSGENIC ANIMALS USING AS EXAMPLES THE *LACI* AND *LACZ* MUTATION MODELS 10

   3.1 The foreign gene construct 10
   3.2 Generation of transgenic animals with shuttle vectors: the transfer method 10
   3.3 Strains and species used 12
   3.4 Target or reporter genes — the *lac* operon 12
   3.5 Transgene shuttle vectors 13
   3.6 Transgenic animal models 14
      3.6.1 *lacI* transgenic model — the Big Blue® construct 14
      3.6.2 *lacZ* transgenic mouse model — the Muta™Mouse construct 15

4. THE TRANSGENIC MUTAGENICITY ASSAY — METHODOLOGY USING AS EXAMPLES *LACI* AND *LACZ* MUTATION MODELS 16

4.1 Treatment schemes 16
4.2 Collection of (target) tissues and isolation of DNA 16
4.3 Recovery of the DNA construct of the shuttle vector from the genomic DNA and in vitro packaging 16
4.4 Infection into bacteria and plating 18
4.5 Detection and quantification of mutations 18
4.5.1 Big Blue® system 18
4.5.2 Muta™Mouse system 19

5. FURTHER TRANSGENIC MUTAGENICITY ASSAYS 21

5.1 *λcII* assay (Big Blue® and Muta™Mouse) 21
5.2 *gpt* delta model 24
5.2.1 *gpt* delta rodents 24
5.2.2 6-TG selection 24
5.2.3 Spi⁻ selection 24
5.3 The *lacZ* plasmid model 26
5.4 Upcoming transgenic models for which no extensive data are available 27
5.4.1 ΦX174 transgenic mouse model 27
5.4.2 λsupF transgenic mouse 28
5.4.3 pKZ1 transgenic recombination model 28
5.4.4 *rpsL* transgenic mouse model 29

6. PARAMETERS AND CRITERIA FOR VALID EXPERIMENTAL DESIGN 30

6.1 Determinants studied using transgenic mutation assays 30
6.1.1 Types of mutations 30
6.1.2 Overall mutant/mutation frequency 30
6.1.3 Spontaneous mutant frequency 31
6.1.4 Sequence analysis 31
6.2 Criteria for valid experimental design of transgenic mutation assays 32
6.2.1 Selection of the most suitable model 32
6.2.2 Duration of exposure 32
6.2.3 Selection of the dose 33

6.2.4 Post-treatment manifestation time 33
6.2.5 Significance of a negative result 35
6.2.6 Factors to consider when comparing the performance of mutation assays 35
6.2.7 Positive control 36
6.2.8 Sensitivity 36
6.2.9 Statistics 37
6.2.10 Analysis of DNA sequence 37
6.2.11 Use of transgenic assays in the detection of gene mutations in germ cells 38

PART II: COMPARISON OF THE *LACI* MODEL AND THE *LACZ* MODEL WITH CONVENTIONAL TEST SYSTEMS 39

7. INTRODUCTION TO PART II 40

7.1 Aim of the comparison and criteria for the selection of data 40
7.2 Choice and limitations of data 41
7.3 Validity of data on transgenic animal mutation assays 42
7.4 Criteria for predictivity of transgenic assays 43

8. COMPARISON OF THE MUTA™MOUSE AND BIG BLUE® ASSAYS 44

9. TRANSGENIC ASSAYS — COMPARISON WITH OTHER ASSAYS 46

9.1 The Muta™Mouse assay and the Big Blue® mouse assay versus the mouse bone marrow micronucleus test 46
9.1.1 The mouse bone marrow micronucleus test: principles and procedures 61
9.1.2 Comparison of data from the mouse bone marrow micronucleus test and transgenic mouse test 61
9.1.2.1 Bone marrow 61
9.1.2.2 All organs 63
9.1.3 Predictivity of the transgenic animal mutagenicity assays and the mouse bone marrow micronucleus test for carcinogenicity 63

9.1.4 Comparison of both test systems 66
9.1.5 Conclusions 68
9.2 The Muta™Mouse assay and the Big Blue® mouse or rat assay versus assays using endogenous reporter genes 70
9.2.1 Results in the mouse spot test compared with those from transgenic animals 70
9.2.1.1 Description of the mouse spot test 71
9.2.1.2 Comparison of the mouse spot test with transgenic mouse model systems 71
9.2.1.3 Predictivity of the transgenic animal mutagenicity assays and the mouse spot test for carcinogenicity 78
9.2.1.4 Advantages and disadvantages of both test systems 79
9.2.1.5 Conclusions 80
9.2.2 Transgenic animal mutagenicity assay versus *Hprt* and other endogenous genes 82
9.2.2.1 Description of endogenous gene animal models 87
9.2.2.2 Comparative studies 89
9.2.2.3 Conclusion 93
9.3 Transgenic animal mutagenicity assays and indirect measure of DNA damage using UDS in vivo assay 94
9.4 Results of transgenic animal mutagenicity assays compared with results of genotoxicity assays in vitro 94
9.4.1 Gene mutation 95
9.4.2 Chromosomal aberration 95
9.4.3 Direct or indirect measure of DNA damage 103
9.4.4 Conclusion 103

10. TRANSGENIC ASSAYS AND CARCINOGENICITY TESTING 104

10.1 Comparison of target organs in carcinogenicity studies with target organs in transgenic animal mutation assays 104
10.1.1 Pattern of target organs 104
10.1.2 Analysis of the predictivity for the liver as target organ 121
10.1.3 Conclusion 127

10.2 Comparison of results of carcinogenicity studies with results from transgenic animal mutagenicity assays 128
10.2.1 Non-genotoxic carcinogens 139
10.2.2 Validity of data on transgenic animal mutation assays 139
10.2.3 Evaluation of the predictivity for carcinogenicity in mice 140
10.2.4 Conclusion 148

PART III: APPLICATIONS OF TRANSGENIC ANIMAL MUTAGENICITY STUDIES 149

11. MUTATION FREQUENCIES AND SEQUENCING DATA AND APPLICATIONS OF THIS INFORMATION IN MECHANISTIC STUDIES 150

11.1 Studies on spontaneous mutant/mutation frequencies (in organs of non-exposed transgenic animals) 150
11.1.1 Sources of spontaneous mutations 150
11.1.2 Spontaneous mutation data: sequence data in organs of non-exposed transgenic animals 151
11.1.3 The frequency and nature of spontaneous mutations versus age in multiple tissues 152
11.2 Examination of fundamental paradigms in genetic toxicology 154
11.2.1 Dose–response relationships 155
11.2.2 Correlation of dose with mutation frequency and carcinogenicity 157
11.2.3 Relationship between DNA adducts, cell proliferation and gene mutations 158
11.3 Studies into the mechanism of action of mutagenicity/carcinogenicity using sequence data 161
11.3.1 Clonal correction and correction for ex vivo mutations 161
11.3.2 Premutagenic lesions 162
11.3.3 Tissue-specific responses 164
11.3.4 Evaluation of genotoxicants that do not appear to interact with DNA 166
11.3.5 Active components of mixtures 166
11.3.6 Active metabolites 166
11.3.7 Investigations into the mechanisms of deletion mutations in vivo 167

11.4 Importance of the transgenic mutation assays for studies other than genetic toxicology 169

PART IV: EVALUATION, CONCLUSIONS AND RECOMMENDATIONS 171

12. EVALUATION OF THE TRANSGENIC ANIMAL MUTAGENICITY ASSAYS BASED ON THIS REVIEW OF THE CURRENT LITERATURE 172

12.1 Features of the assay 172
12.2 Gene mutation assay — implications for testing 172
12.2.1 Reliability of a negative result 173
12.3 Comparison with endogenous genes 174
12.4 Molecular analysis and mechanistic studies 174
12.5 Animal welfare and economy 175

13. CONCLUSIONS 177

14. RECOMMENDATIONS 178

REFERENCES 179

APPENDIX 1: MASTER TABLE 215

APPENDIX 2: THE *CII* ASSAY IN TRANSGENIC RODENT STUDIES 279

RESUME 287

RESUMEN 293

NOTE TO READERS OF THE CRITERIA MONOGRAPHS

Every effort has been made to present information in the criteria monographs as accurately as possible without unduly delaying their publication. In the interest of all users of the Environmental Health Criteria monographs, readers are requested to communicate any errors that may have occurred to the Director of the International Programme on Chemical Safety, World Health Organization, Geneva, Switzerland, in order that they may be included in corrigenda.

Environmental Health Criteria

PREAMBLE

Objectives

In 1973 the WHO Environmental Health Criteria Programme was initiated with the following objectives:

(i) to assess information on the relationship between exposure to environmental pollutants and human health, and to provide guidelines for setting exposure limits;
(ii) to identify new or potential pollutants;
(iii) to identify gaps in knowledge concerning the health effects of pollutants;
(iv) to promote the harmonization of toxicological and epidemiological methods in order to have internationally comparable results.

The first Environmental Health Criteria (EHC) monograph, on mercury, was published in 1976, and since that time an ever-increasing number of assessments of chemicals and of physical effects have been produced. In addition, many EHC monographs have been devoted to evaluating toxicological methodology, e.g., for genetic, neurotoxic, teratogenic and nephrotoxic effects. Other publications have been concerned with epidemiological guidelines, evaluation of short-term tests for carcinogens, biomarkers, effects on the elderly and so forth.

Since its inauguration the EHC Programme has widened its scope, and the importance of environmental effects, in addition to health effects, has been increasingly emphasized in the total evaluation of chemicals.

The original impetus for the Programme came from World Health Assembly resolutions and the recommendations of the 1972 UN Conference on the Human Environment. Subsequently the work became an integral part of the International Programme on Chemical Safety (IPCS), a cooperative programme of UNEP, ILO and WHO. In this manner, with the strong support of the new partners, the importance of occupational health and environmental effects was fully recognized. The EHC monographs have become widely established, used and recognized throughout the world.

The recommendations of the 1992 UN Conference on Environment and Development and the subsequent establishment of the Intergovernmental Forum on Chemical Safety with the priorities for action in the six programme areas of Chapter 19, Agenda 21, all lend further weight to the need for EHC assessments of the risks of chemicals.

Scope

The criteria monographs are intended to provide critical reviews on the effect on human health and the environment of chemicals and of combinations of chemicals and physical and biological agents. As such, they include and review studies that are of direct relevance for the evaluation. However, they do not describe *every* study carried out. Worldwide data are used and are quoted from original studies, not from abstracts or reviews. Both published and unpublished reports are considered, and it is incumbent on the authors to assess all the articles cited in the references. Preference is always given to published data. Unpublished data are used only when relevant published data are absent or when they are pivotal to the risk assessment. A detailed policy statement is available that describes the procedures used for unpublished proprietary data so that this information can be used in the evaluation without compromising its confidential nature (WHO (1990) Revised Guidelines for the Preparation of Environmental Health Criteria Monographs. PCS/90.69, Geneva, World Health Organization).

In the evaluation of human health risks, sound human data, whenever available, are preferred to animal data. Animal and in vitro studies provide support and are used mainly to supply evidence missing from human studies. It is mandatory that research on human subjects is conducted in full accord with ethical principles, including the provisions of the Helsinki Declaration.

The EHC monographs are intended to assist national and international authorities in making risk assessments and subsequent risk management decisions. They represent a thorough evaluation of risks and are not, in any sense, recommendations for regulation or standard setting. These latter are the exclusive purview of national and regional governments.

Content

The layout of EHC monographs for chemicals is outlined below.

- Summary — a review of the salient facts and the risk evaluation of the chemical
- Identity — physical and chemical properties, analytical methods
- Sources of exposure
- Environmental transport, distribution and transformation
- Environmental levels and human exposure
- Kinetics and metabolism in laboratory animals and humans
- Effects on laboratory mammals and in vitro test systems
- Effects on humans
- Effects on other organisms in the laboratory and field
- Evaluation of human health risks and effects on the environment
- Conclusions and recommendations for protection of human health and the environment
- Further research
- Previous evaluations by international bodies, e.g., IARC, JECFA, JMPR

Selection of chemicals

Since the inception of the EHC Programme, the IPCS has organized meetings of scientists to establish lists of priority chemicals for subsequent evaluation. Such meetings have been held in Ispra, Italy, 1980; Oxford, United Kingdom, 1984; Berlin, Germany, 1987; and North Carolina, USA, 1995. The selection of chemicals has been based on the following criteria: the existence of scientific evidence that the substance presents a hazard to human health and/or the environment; the possible use, persistence, accumulation or degradation of the substance shows that there may be significant human or environmental exposure; the size and nature of populations at risk (both human and other species) and risks for environment; international concern, i.e., the substance is of major interest to several countries; adequate data on the hazards are available.

If an EHC monograph is proposed for a chemical not on the priority list, the IPCS Secretariat consults with the Cooperating Organizations and all the Participating Institutions before embarking on the preparation of the monograph.

Procedures

The order of procedures that result in the publication of an EHC monograph is shown in the flow chart on p. xiv. A designated staff member of IPCS, responsible for the scientific quality of the document, serves as Responsible Officer (RO). The IPCS Editor is responsible for layout and language. The first draft, prepared by consultants or, more usually, staff from an IPCS Participating Institution, is based on extensive literature searches from reference databases such as Medline and Toxline.

The draft document, when received by the RO, may require an initial review by a small panel of experts to determine its scientific quality and objectivity. Once the RO finds the document acceptable as a first draft, it is distributed, in its unedited form, to well over 150 EHC contact points throughout the world who are asked to comment on its completeness and accuracy and, where necessary, provide additional material. The contact points, usually designated by governments, may be Participating Institutions, IPCS Focal Points or individual scientists known for their particular expertise. Generally some four months are allowed before the comments are considered by the RO and author(s). A second draft incorporating comments received and approved by the Director, IPCS, is then distributed to Task Group members, who carry out the peer review, at least six weeks before their meeting.

The Task Group members serve as individual scientists, not as representatives of any organization, government or industry. Their function is to evaluate the accuracy, significance and relevance of the information in the document and to assess the health and environmental risks from exposure to the chemical. A summary and recommendations for further research and improved safety aspects are also required. The composition of the Task Group is dictated

EHC PREPARATION FLOW CHART

Commitment to draft EHC

Document preparation initiated

Revision as necessary

Draft sent to IPCS Responsible Officer (RO)

Possible meeting of a few experts to resolve controversial issues

Responsible Officer, Editor check for coherence of text and readability (not language editing)

First Draft

International circulation to Contact Points (150+)

Comments to IPCS (RO)

Review of comments, reference cross-check; preparation of Task Group (TG) draft

Editor

Task Group meeting

Working group, if required

Insertion of TG changes

Post-TG draft; detailed reference cross-check

Editing

Graphics

Word-processing

French/Spanish translations of Summary

Camera-ready copy

Library for CIP Data

Final editing

Approval by Director, IPCS

WHO Publication Office

Printer

Proofs

Publication

routine procedure

optional procedure

by the range of expertise required for the subject of the meeting and by the need for a balanced geographical distribution.

The three cooperating organizations of the IPCS recognize the important role played by nongovernmental organizations. Representatives from relevant national and international associations may be invited to join the Task Group as observers. Although observers may provide a valuable contribution to the process, they can speak only at the invitation of the Chairperson. Observers do not participate in the final evaluation of the chemical; this is the sole responsibility of the Task Group members. When the Task Group considers it to be appropriate, it may meet in camera.

All individuals who as authors, consultants or advisers participate in the preparation of the EHC monograph must, in addition to serving in their personal capacity as scientists, inform the RO if at any time a conflict of interest, whether actual or potential, could be perceived in their work. They are required to sign a conflict of interest statement. Such a procedure ensures the transparency and probity of the process.

When the Task Group has completed its review and the RO is satisfied as to the scientific correctness and completeness of the document, it then goes for language editing, reference checking and preparation of camera-ready copy. After approval by the Director, IPCS, the monograph is submitted to the WHO Office of Publications for printing. At this time a copy of the final draft is sent to the Chairperson and Rapporteur of the Task Group to check for any errors.

It is accepted that the following criteria should initiate the updating of an EHC monograph: new data are available that would substantially change the evaluation; there is public concern for health or environmental effects of the agent because of greater exposure; an appreciable time period has elapsed since the last evaluation.

All Participating Institutions are informed, through the EHC progress report, of the authors and institutions proposed for the drafting of the documents. A comprehensive file of all comments received on drafts of each EHC monograph is maintained and is available on request. The Chairpersons of Task Groups are briefed before each meeting on their role and responsibility in ensuring that these rules are followed.

WHO TASK GROUP ON ENVIRONMENTAL HEALTH CRITERIA FOR TRANSGENIC ANIMAL MUTAGENICITY TESTING

The first draft of the EHC monograph was prepared for IPCS by the Fraunhofer Institute, Hanover, Germany, in 2004. It was widely distributed by IPCS for international peer review in late July to early August, with comments due by 1 October 2004. A revised proposed draft document, taking into account comments received, was prepared by the Fraunhofer Institute. An EHC Task Group was convened on 13–17 December 2004, in Hanover, Germany, to further develop and finalize the document.

Ms C. Vickers was responsible for the overall scientific content of the monograph.

The efforts of all who helped in the preparation and finalization of the monograph are gratefully acknowledged.

* * *

Risk assessment activities of the International Programme on Chemical Safety are supported financially by the Department of Health and Department for Environment, Food & Rural Affairs, United Kingdom; Environmental Protection Agency, Food and Drug Administration, and National Institute of Environmental Health Sciences, USA; European Commission; German Federal Ministry of Environment, Nature Conservation and Nuclear Safety; Health Canada; Japanese Ministry of Health, Labour and Welfare; and Swiss Agency for Environment, Forests and Landscape.

Task Group Members

Dr T. Chen, National Center for Toxicological Research, Food and Drug Administration, Jefferson, Arkansas, USA

Dr S. Dean, Huntingdon Life Sciences, Huntingdon, Cambridgeshire, United Kingdom

Dr G.R. Douglas, Healthy Environments and Consumer Safety Branch, Health Canada, Ottawa, Ontario, Canada

Dr J.A. Heddle, Department of Biology, York University, Toronto, Ontario, Canada (*Chairperson*)

Dr I.B. Lambert, Department of Biology, Carleton University, Ottawa, Ontario, Canada

Dr T. Nohmi, National Institute of Health Sciences, Tokyo, Japan

Dr P.J. Sykes, Department of Haematology and Genetic Pathology, Flinders University and Medical Centre, Bedford Park, South Australia

Dr V. Thybaud, Paris Research Center, Vitry sur Seine, France

Dr C.R. Valentine, National Center for Toxicological Research, Food and Drug Administration, Jefferson, Arkansas, USA (*Co-Rapporteur*)

Dr J. van Benthem, National Institute of Public Health and the Environment (RIVM), Bilthoven, The Netherlands

Secretariat

Dr A. Bitsch, Fraunhofer Institute of Toxicology and Experimental Medicine, Hanover, Germany

Dr J. Kielhorn, Fraunhofer Institute of Toxicology and Experimental Medicine, Hanover, Germany (*Co-Rapporteur*)

Dr I. Mangelsdorf, Fraunhofer Institute of Toxicology and Experimental Medicine, Hanover, Germany

Ms C. Vickers, International Programme on Chemical Safety, World Health Organization, Geneva, Switzerland

Dr U. Wahnschaffe, Fraunhofer Institute of Toxicology and Experimental Medicine, Hanover, Germany

ACRONYMS AND ABBREVIATIONS

A	adenine
Aprt	adenine phosphoribosyl transferase
bp	base pair
C	cytosine
CCRIS	Chemical Carcinogenesis Research Information System
cDNA	complementary deoxyribonucleic acid
CICAD	Concise International Chemical Assessment Document
CpG	cytosine and guanine connected by a phosphodiester bond
dA	deoxyadenosine
dG	deoxyguanosine
Dlb-1	Dolichos biflorus-1
DMBA	7,12-dimethylbenz[*a*]anthracene
DMDBC	5,9-dimethyldibenzo[*c,g*]carbazole
DNA	deoxyribonucleic acid
EHC	Environmental Health Criteria
FAO	Food and Agriculture Organization of the United Nations
G	guanine
GENE-TOX	Genetic Toxicology (data bank)
Hprt	hypoxanthine-guanine phosphoribosyltransferase
HSDB	Hazardous Substances Data Bank
IARC	International Agency for Research on Cancer
ILO	International Labour Organization
i.p.	intraperitoneal
IPCS	International Programme on Chemical Safety
IRIS	Integrated Risk Information System
IWGT	International Workshop on Genotoxicity Testing

JECFA	Joint FAO/WHO Expert Committee on Food Additives
JMPR	Joint FAO/WHO Meeting on Pesticide Residues
kb	kilobase (1000 base pairs)
LOH	loss of heterozygosity
MAK	German Commission for the Investigation of Health Hazards of Chemical Compounds in the Work Area (MAK Commission), which performs critical data evaluation for MAK (maximum workplace concentration) values and classification of carcinogens
Mb	megabase (1 000 000 base pairs)
MTD	maximum tolerated dose
NOEL	no-observed-effect level
NTP	National Toxicology Program (USA)
OECD	Organisation for Economic Co-operation and Development
Oua	ouabain
PCE	polychromatic erythrocyte
P-gal	phenyl-β-D-galactopyranoside
RNA	ribonucleic acid
RO	Responsible Officer
SBA	single burst analysis
s.c.	subcutaneous
SCE	sister chromatid exchange
SICR	somatic intrachromosomal recombination
Spi^-	sensitive to P2 interference
T	thymine
6-TG	6-thioguanine
Tk	thymidine kinase
tRNA	transfer ribonucleic acid
UDP	uridine diphosphate

UDS	unscheduled DNA synthesis
UN	United Nations
UNEP	United Nations Environment Programme
USA	United States of America
UVB	ultraviolet B
WHO	World Health Organization
X-gal	5-bromo-4-chloro-3-indolyl-β-D-galactopyranoside

GLOSSARY

Allosteric protein	Protein that changes from one conformation to another when it binds to another molecule or when it is covalently modified. This change in conformation alters the activity of the protein.
Aneugenic, aneuploidy	Is used for agents giving rise to numerical chromosomal aberrations in cells or organisms.
Clastogenic	Is used for agents giving rise to structural chromosomal aberrations in populations of cells or organisms.
Clonal expansion	Increase in the number of mutants from a single mutated cell. These mutants will have identical sequence changes. Clonal expansion can increase the mutant frequency if it occurs in early development of organ or tissue (jackpot mutation).
Coding region	Region of DNA that is translated into a protein.
Cos site	Cohesive ends of the λ genome.
Deletion	Loss of adjacent bases in DNA. Small deletions may remove one or a few base pairs within a gene, while larger deletions can remove an entire gene or several neighbouring genes. The deleted DNA may alter the function of the resulting protein(s).
Duplication	Insertion of a DNA sequence corresponding to an existing sequence. This type of mutation may alter the function of the resulting protein if it occurs within a gene.
Frameshift	The addition or loss of DNA bases within a gene such that it changes the reading frame. A reading frame consists of groups of three bases that each code for one amino acid. A frameshift mutation shifts the grouping of these bases and changes the code for amino acids. The resulting protein is usually non-functional. Insertions, deletions and duplications can all be frameshift mutations.

Genotoxic	Able to alter the structure, information content or segregation of DNA.
Head to tail	A succession of vectors in the same 3' → 5' direction without inversions.
Hemizygous	Having only one member of a chromosome pair or chromosome segment rather than the usual two; refers in particular to X-linked genes in males who under normal circumstances have only one X chromosome.
Insertion	Addition of DNA bases in a gene. As a result, the protein made by the gene may not function properly.
Jackpot mutation	Exceptionally high mutant frequency that is due to clonal expansion of a single mutant.
Lysogeny	Integration of the phage gene into the *Escherichia coli* genome without further replication and viral synthesis.
Manifestation time	The time between the exposure and collection of organs or tissues. This manifestation or fixation time is required to fix the DNA damage into an irreversible mutation and for the cells of the tissue to be largely replaced after exposure.
Missense mutation	This type of mutation is a change in one DNA base pair that results in the substitution of one amino acid for another in the protein made by a gene.
Mutagenic	Capable of giving rise to mutations.
Mutant frequency	The ratio of the number of mutant plaques or colonies to the total number of plaques tested (which is normally estimated from titre plates).
Mutation	A permanent structural alteration in DNA. In most cases, DNA changes either have no effect or cause harm, but occasionally a mutation can improve an organism's chance of surviving and passing the beneficial change on to its descendants.

Mutation frequency	The mutant frequency corrected for clonal expansion.
Mutation spectrum	The relative frequencies of different types of mutations and the pattern of their occurrence within a DNA sequence.
Non-genotoxic carcinogen	Carcinogen whose primary action does not involve DNA alterations.
Nonsense mutation	A nonsense mutation is also a change in one DNA base pair. Instead of substituting one amino acid for another, however, the altered DNA sequence prematurely signals the cell to stop building a protein. This type of mutation results in a shortened protein that may function improperly or not at all.
Operator	Short region of DNA in a (prokaryotic) chromosome that controls the transcription of an adjacent gene.
Operon	A functional unit of transcription consisting of one or more structural genes and two associated segments of DNA: an operator (the switch) and a promoter (a binding site for the transcription enzyme). Operons occur primarily in prokaryotes.
Plaque	A clear area on a bacterial lawn, left by lysis of the bacteria through progressive infections by a phage.
Plasmid	A circular piece of DNA that exists apart from the chromosome. Plasmids (vectors) are often used in genetic engineering to carry desired genes into organisms.
Point mutation	A mutation that changes a single DNA base pair of a gene.
Promotor	The normal loading point for RNA polymerase, often the point at which transcription is initiated.

Reporter gene	A gene whose phenotypic expression is easy to monitor; reporter genes are "markers" widely used for analysis of mutationally altered genes as well as gene regulation.
Repressor	A protein that regulates a gene by turning it off.
Target organ	In transgenic animal mutation systems, the target organ is the organ of a transgenic animal in which mutagenic effects (increased mutation frequency) were detected after exposure to the test substance. Target organs for carcinogenicity are those organs where tumours arise.
Transgenic	An experimentally produced organism in which DNA has been artificially introduced and incorporated into the organism's germline, usually by injecting the foreign DNA into the nucleus of a fertilized embryo.
Transition	A base pair substitution in which the orientation of the purine and pyrimidine bases on each DNA strand remains the same; i.e. A:T→G:C, T:A→G:C.
Transversion	A base pair substitution in which the purine–pyrimidine orientation on each DNA strand is reversed; i.e. A:T→T:A.
Vector	An agent, such as a virus or a plasmid, that carries a modified or foreign gene. When used in gene therapy, a vector delivers the desired gene to a target cell.

1. SUMMARY

The aim of this document is to introduce newcomers in this field to transgenic mutagenicity assays and to assess the possible role of these assays in toxicology testing and mechanistic research.

A transgenic animal carries foreign DNA that is integrated into the chromosomal DNA of the animal and is present in all cells. In transgenic mutagenicity assays, the foreign DNA is an exogenous gene (transgene) injected into the nucleus of a fertilized rodent embryo. These reporter genes are transmitted by the germ cells and thus are present in all cells of the newborn rodent and can be used to detect mutation frequency.

Part I of this document (Chapters 2–6) gives a short overview of in vivo genotoxicity testing. The methods employed in the design of transgenic animals are explained, giving details of the DNA construct and of the methods used for inserting the construct into the recipient animals. As examples, transgenic models — in particular the *lacI* model, commercially available as the Big Blue® mouse and Big Blue® rat, and the *lacZ* model, commercially available as the Muta™Mouse — are described, as well as more recently developed models, such as *λcII*, the *gpt* delta, *lacZ* plasmid and ΦX174.

Study design is critical to the validity of a study for determining positive/negative mutagenicity of a test compound. The choice of the mutagenic target gene, species and tissue should be based on any prior knowledge of the pharmacological/toxicological parameters of the test agent. Since the selection of dose, dosing schedule and post-treatment sampling time varies for the optimal detection of mutation frequency for different tissues and agents, a protocol has been recommended that optimizes detection of all mutagens, regardless of potency or target tissue. A negative result obtained using a robust protocol should be considered as valid.

Part II (Chapters 7–10) gives an overview of data published on chemicals tested using the *lacI* model and the *lacZ* model, compares these with data available with conventional systems and discusses the outcomes. These models were chosen because they are the only two systems with enough data available to allow comparisons and analyses to be made.

The limited data available suggest that there is significant agreement with respect to the results obtained with the Muta™Mouse and the Big Blue® mouse or rat assay. Any observed differences between the Muta™Mouse and the Big Blue® mouse assay are likely to be attributable to the different experimental design used in the particular studies, rather than differences in the sensitivity of the transgenic reporter genes per se.

The results of the transgenic mutation assays were compared with those of the mouse bone marrow micronucleus assay for 44 substances. Although the majority of the results were often similar, as many of the chemicals tested were potent carcinogens, the assays were complementary — that is, there was a significant improvement in the detection of carcinogens when both assays were used. The theoretical advantage of using two assays that detect different genotoxic end-points seems to be confirmed by this result. The ability of the transgenic animal assays to detect gene mutations in multiple tissues is also a distinct advantage.

Although the mouse spot test is a standard genotoxicity test system according to Organisation for Economic Co-operation and Development (OECD) guidelines, this system has seldom been used for detection of somatic mutations in vivo in recent decades. The results of a comparison of both systems in this document showed that the transgenic mouse assay has several advantages over the mouse spot test and is a suitable test system to replace the mouse spot test for detection of gene but not chromosome mutations in vivo.

Despite differences in the mutational properties of the various model mutagens, the responses of the exogenous loci (*lacI*, *lacZ* transgene) and the endogenous loci (*Dlb-1*, *Hprt*) were generally qualitatively similar following acute treatments. Several studies suggest that the lower somatic mutant frequency in the endogenous genes may provide enhanced sensitivity under such conditions. However, comparisons of transgenes and endogenous genes are difficult because of differences between the optimal experimental protocols for the different types of genes; in the neutral transgenes, sensitivity for the detection of mutations is increased with the longer administration times that are currently recommended.

The limited data comparing unscheduled DNA synthesis (UDS) with *lacI* and *lacZ* suggest that transgenic animal assays exhibit superior predictivity compared with the UDS test, which measures DNA damage. Results from transgenic animal assays (*lacI* and *lacZ*) with over 50 chemicals agreed with results from in vitro data on gene mutation, chromosomal aberration and direct or indirect measures for DNA damage by these chemicals. A major advantage of the transgenic mouse/rat mutation assay compared with other in vivo mutagenicity tests is that mutagenic events in any organ can be detected. Therefore, an analysis was made to determine whether target organs in carcinogenicity studies can be predicted by transgenic mutation assays. In most cases, mutations were found in the target organs of the carcinogenicity studies. For several presumed genotoxic carcinogens, organs investigated in the transgenic mutagenicity assays, which were not target organs in carcinogenicity studies, were positive. As this has occurred for several compounds, it is unlikely to be explained by insufficient specificity with regard to target organs for carcinogenicity. Instead, it leads to the conclusion that genotoxicity is expressed in several organs in the body and that tumours do not develop in all these organs due to other factors. Carcinogens with a presumed non-genotoxic mode of action generally produce negative results in the transgenic animal assays. Very few data are available on substances that gave negative results in carcinogenicity assays on mice. However, for these few non-carcinogens, the results in transgenic mice were also negative. The available data suggest that the sensitivity and positive predictivity of the transgenic assays for carcinogenicity are high.

Part III (Chapter 11) describes studies in which transgenic mutation assays (in particular, the *lacI* and *lacZ* model using *cII* and the *gpt* delta rodent system) have been used as mechanistic research tools. Due to the ease of sequencing the *cII* gene for mutational spectra, it is increasingly used instead of *lacI* and *lacZ* in the Muta™Mouse and Big Blue® models for sequencing studies. The *gpt* delta model is also used because of the ease of sequencing and, especially, because it detects deletions much larger than those detected by all but the *lacZ* plasmid assay.

Spontaneous mutations have been studied in almost all transgenic animal mutagenicity assays: *lacZ*, *lacI* and *cII*, *lacZ* plasmid and *gpt* delta mice. In all systems, the predominant type of spontaneous mutation is G:C→A:T transitions, with most occurring at 5'-CpG sites, suggesting that the deamination of 5-methylcytosine is the main mechanism of mutagenesis.

The frequency and nature of spontaneous mutations have been studied. The factors that affect the inferred mutation rate are site of integration of the transgene, age, tissue and strain. About half of all mutations arise during development (and half of these in utero). Several studies have examined the frequency and nature of spontaneous mutations versus age in multiple tissues and found that, with the exception of studies in the plasmid mouse, the spectrum of mutation types was similar with age and tissue type in adult animals. It did not vary with differences in gender or mouse genetic background. The mutation frequency in the male germline was consistently the lowest, remaining essentially unchanged in old age.

Transgenic animal assays have been found to be useful tools in the examination of fundamental paradigms in genetic toxicology. Recent studies using these systems have addressed the issues of 1) dose–response relationship of genotoxic carcinogens and 2) the relationships among DNA adduct formation, mutation frequency and cancer in rodents. Further important application of these transgenic rodent assays has been in fundamental studies on the origin of mutations and the roles of various biological processes in preventing them. These studies have included studies of DNA repair mechanisms, carcinogenesis, ageing and inherited genetic conditions affecting these processes.

While mutation spectra from DNA sequence data are not considered mandatory for the evaluation of gene mutation in vivo in the case of clear positive or negative results, they are useful for factors relating to the mechanism of mutagenesis. The ability to sequence induced mutations in transgenic reporter genes provides an investigator with important information regarding several aspects of mutation. Examples are given of studies that demonstrate how transgenic animal assays and subsequent spectral analysis can be used to examine different aspects of the activity of mutagenic agents: for example, 1) clonal correction and correction for ex vivo mutations,

2) premutagenic lesions, 3) tissue-specific responses, 4) evaluation of genotoxicants that do not interact with DNA, 5) determination of the active components of mixtures, 6) determinations of active metabolites and 7) investigations into the mechanisms of deletion mutations in vivo.

Part IV (Chapters 12–14) evaluates the role and potential added value of transgenic mutation assays in toxicology and risk assessment. To date, transgenic mutagenicity assays have not been heavily used by industry in toxicological screening, in large part because an OECD Test Guideline has not yet been developed. Recently, an internationally harmonized protocol has been recommended (Thybaud et al., 2003), and this protocol should form the basis for such a guideline.

The IPCS Task Group recommends the development of such a guideline. The utility of such a guideline is based, in part, on the fact that the transgenic animal assays are capable of detecting gene mutations. If such a protocol is used, a negative result can be considered as reliable.

The IPCS Task Group further recommends that transgenic mutagenicity assays be included in the IPCS Qualitative Scheme for Mutagenicity and other testing strategies.

For future research, the IPCS Task Group recommends the testing of a number of well established non-carcinogens according to a robust protocol (e.g. Thybaud et al., 2003). Transgenic mutagenicity assays should be recommended as tools for studies of the mechanistic relationship between mutation and carcinogenesis and for studies of germline mutagenesis.

PART I:

OVERVIEW OF GENOTOXICITY TESTING AND TRANSGENIC ANIMAL MUTAGENICITY SYSTEMS

2. OVERVIEW OF GENOTOXICITY TESTING

The potential genotoxicity of chemicals is assessed in short-term in vitro and in vivo genotoxicity tests. Under in vitro conditions, there are sufficient assays for detecting both gene mutations (e.g. Ames test in bacteria) and chromosomal aberrations in mammalian cells.

However, testing under in vivo conditions is essential to confirm in vitro tests, as it is impossible to mimic in vitro whole animal processes such as absorption, tissue distribution, metabolism and excretion of the chemical and its metabolites. This lack of a well validated in vivo gene mutation test hinders the assessment of the genotoxic potential of chemicals.

2.1 Gene mutation assays in vivo using endogenous genes

Only a few mutation assays are available for endogenous genes in mammalian cells (e.g. *Hprt*, *Aprt*, *Tk* or *Dlb-1*). Moreover, the determination of the mutation frequency in these assays is restricted to only a few tissues. Determination of the mutation frequency of *Dlb-1* is restricted to the small intestine and possibly the colon; determination of the mutation frequency of *Hprt*, *Aprt* and *Tk* is restricted to those tissues that express the reporter genes and that can be subcultured in vitro. A short description of the mutation assays for these genes is given in section 9.2.2.1.

2.2 Gene mutation assays in vivo using transgenes

These assays are performed in transgenic animals — that is, animals that possess an exogenous reporter gene, a so-called transgene (e.g. *lacZ* or *lacI*). Based on the shuttle vector used, there are two main approaches for the use of transgenic rodent models for mutagenicity testing: 1) using a transgene in a bacteriophage vector and 2) using a transgene in a plasmid vector. A short overview of in vivo genotoxicity assays, including transgenes, is given in Table 1, and data on transgenic gene mutation assays are given in chapters 3, 4 and 5.

Table 1. Assays for testing genotoxicity *in vivo*[a]

Direct and indirect measures of DNA damage	Gene mutation assays	Chromosomal aberration and/or aneuploidy assays
	Assays using endogenous reporter genes	
In vivo sister chromatid exchange in rodents (no guideline)	**Mouse spot test** (OECD 484: OECD, 1986a)	**Mouse bone marrow micronucleus test** (OECD 474: OECD, 1997a)
Unscheduled DNA synthesis in rodents (OECD 486: OECD, 1971)	Sex-linked recessive lethal test in *Drosophila melanogaster* (OECD 477: OECD, 1984a)	Mammalian bone marrow chromosomal aberration test (OECD 475: OECD, 1997b)
	***Aprt* mouse** ***Tk* mouse** ***Hprt* somatic mutation assay** ***Dlb-1* specific locus assay** (no guidelines)	Rodent dominant lethal assay (OECD 478: OECD, 1984b)
		Mammalian germ cell cytogenetic assay (OECD 483: OECD, 1997c)
		Mouse heritable translocation assay (OECD 485: OECD, 1986b)
	Assays using transgenic reporter genes (no OECD guidelines at present)	
Single cell gel/ comet assay in rodents[b]	**Big Blue® mouse and rat (*lacI* and *cII*)** **Muta™Mouse (*lacZ* and *cII*)** **ΦX174*E/A* mouse** ***gpt* delta mouse and rat** ***lacZ* plasmid mouse φmodel** ***rpsL* transgenic mouse** **λsupF transgenic mouse**	**pKZ1 transgenic recombination mutagenesis assay**

[a] Bold signifies assays discussed in this document.
[b] Tice et al. (2000); no OECD guideline available at present.

3. CONSTRUCTION/PRODUCTION OF TRANSGENIC ANIMALS USING AS EXAMPLES THE *LACI* AND *LACZ* MUTATION MODELS

A transgenic animal carries foreign DNA that is integrated into the chromosomal DNA of the animal in all cells. This chapter describes the methods employed in the design of transgenic animals using the examples of Big Blue® (Kohler et al., 1991a; Stratagene, 2002) and Muta™Mouse (Vijg & Douglas, 1996), giving details of the DNA construct and of the methods used for inserting the construct into the recipient animals. Following this, details of some other transgenic systems — for example, the *λcII* assay, the *gpt* delta rodent system and Spi⁻ selection, the *lacZ* plasmid mouse and ΦX174*E/A* — are introduced.

3.1 The foreign gene construct

There are two essential parts of the foreign gene construct or *transgene* in currently used transgenic mutation test systems: 1) a shuttle vector for recovering the target gene DNA from the tissue of the transgenic animal; and 2) the *reporter gene*, which may serve concurrently as a *target gene* for scoring mutations. The transgene is constructed using recombinant DNA technologies.

3.2 Generation of transgenic animals with shuttle vectors: the transfer method

Big Blue® animals and the Muta™Mouse have been produced by pronucleus microinjection (Fig. 1), a technique that is currently the most successful and most widely used method of producing transgenic animals. The method allows an early integration of the transgene into the host DNA, which is important to ensure that transgenic DNA is apparent in all cells of the host.

Pronucleus microinjection was first described by Gordon and colleagues (Gordon et al., 1980; Gordon & Ruddle, 1983). Male and female pronuclei are microscopically visible several hours following the entry of the sperm into the oocyte. The transgene may be micro-injected into either of these pronuclei, with equivalent results.

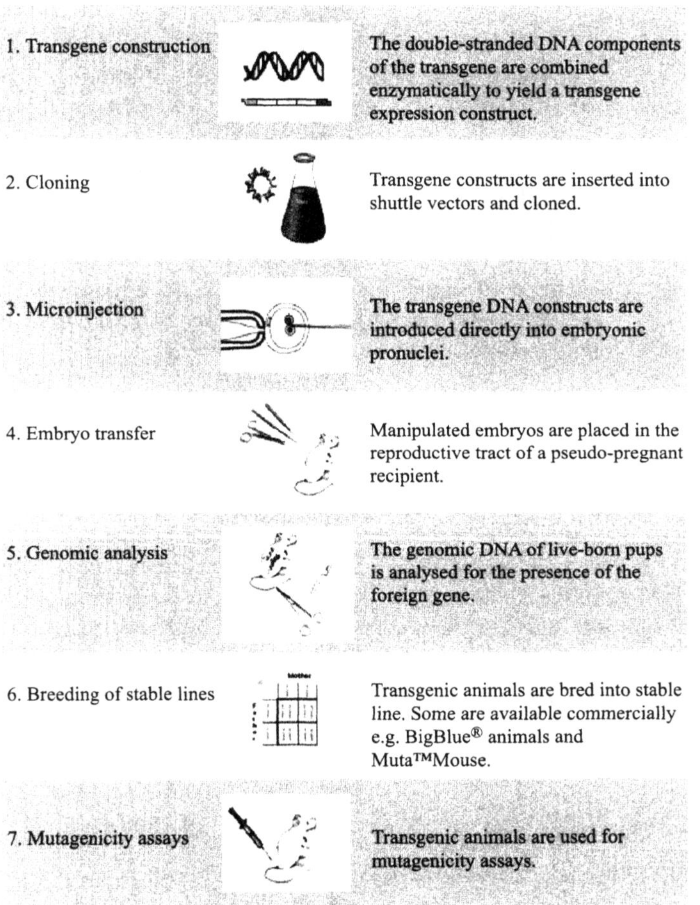

Fig. 1. Schematic presentation of transgene construction and preparation of transgenic animals.

Once a group of pronuclei has been injected with transgene DNA, the eggs are transferred in medium for incubation and visual evaluation within the next few hours. All embryos that are visually graded as viable are then transferred to a recipient female oviduct (Fig. 1).

The offspring are born with several copies of this new information in every cell. Each animal that develops after receiving the transgene DNA is referred to as the founder (F0) of a new transgenic lineage. If the germ cells of the founder transmit the transgene stably, then all descendants of this animal are members of a unique transgenic lineage. The new transgenic locus is present in only one of the two chromosomes; therefore, the genotype of the founder is described as hemizygous for the transgene. A homozygous genotype, in which a pair of transgene alleles is present, may be produced by the mating of a pair of hemizygous F1 siblings.

3.3 Strains and species used

For the construction of a Big Blue® mouse, fertilized eggs of C57BL/6 mice were microinjected with the foreign DNA construct (Kohler et al., 1991a). A founder mouse was crossed with the non-transgenic C57BL/6, and F1 offspring were used in subsequent experiments. In another Big Blue® hybrid, the founder line A1 derived from the C57BL/6 crossed with the C3H line is used to produce the same genetic background as the United States National Toxicology Program (NTP) bioassay test mice B6C3F1 (Kohler et al., 1991a).

A transgenic Big Blue® rat line has been developed in F344 rats (Dycaico et al., 1994; Gollapudi et al., 1998).

For the construction of Muta™Mouse, fertilized eggs of CD-2 F1 (BALB/c × DBA/2) were microinjected with 150 copies of the monomeric λgt10LacZ vector. Four mice with different copy numbers were selected to be bred into strains. The Muta™Mouse is strain 40.6 (Gossen et al., 1989).

3.4 Target or reporter genes — the *lac* operon

Several target genes for mutations are currently used in genotoxicity testing in mammalian models. From these, the bacterial *lacI* and *lacZ* genes have been studied in most detail. Both genes are involved in lactose metabolism of *Escherichia coli*.

In particular, the *lacI* gene has been used for many years as a convenient mutagenic target (e.g. Schaaper et al., 1986, 1990; Horsfall & Glickman, 1989). Both *lacI* and *lacZ* genes are part of the lac operon (Fig. 2). The *lacI* gene is located directly upstream of the lac promoter and encodes the lac repressor, which suppresses *lacZ* transcription (Gilbert & Müller-Hill, 1967). This lac repressor comprises four identical polypeptides ("homotetramer"). One part of the molecule is able to recognize and bind to the 24 base pairs (bp) of the operator region of the lac operon structural genes, thereby suppressing transcription of these genes. Another part of the repressor contains sites that bind to lactose or related molecules; lactose binding causes allosteric changes in tetramer conformation that prevent binding to the operator. This enables *lacZ* transcription. The *lacZ* gene codes for the enzyme β-galactosidase (for *lacZ* gene sequence, see Kalnins et al., 1983), which is the reporter gene, producing blue colour in the presence of certain substrates.

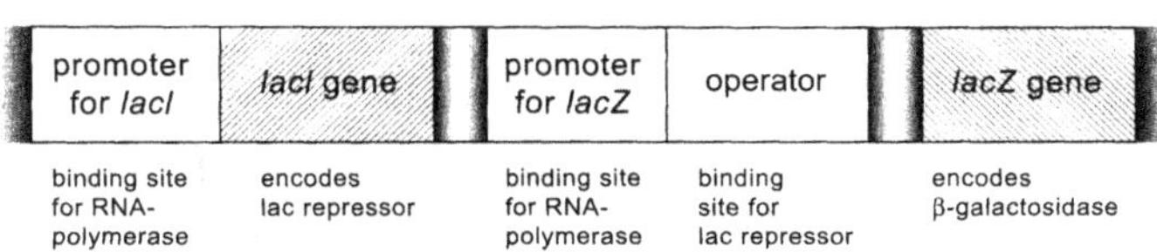

Fig. 2. Simplified scheme of the *lac* operon.

3.5 Transgene shuttle vectors

Recombinant DNA must be taken up by the host cells and incorporated into the genome for replication; it must then be recovered in a form that will replicate and express the reporter gene in *E. coli* for mutation scoring. This is achieved by incorporating the DNA construct in either a bacteriophage vector (e.g. the λ bacteriophages) or a plasmid vector (e.g. pUR288 shuttle plasmid vector; see section 5.3). Plasmid vectors are not dealt with in detail in this document.

The use of λ bacteriophage as a shuttle vector was first developed in mouse fibroblasts by Glazer et al. (1986) and was applied to transgenic mice by Gossen et al. (1989). The *E. coli* λ phage has a genome size of about 49 kilobases (kb) and can be used to carry DNA fragments limited to a maximum of 15 kb. The Big Blue® and Muta™Mouse assays use transgenic mice harbouring chromosomally integrated λ bacteriophage containing mutational target genes. The transgene is integrated within the mouse genome in tandemly repeated vectors. In these models, targets for mutations in the λ vectors are the *E. coli lacI* gene (Kohler et al., 1991a, 1991b), the *lacZ* gene (Gossen et al., 1989; Gossen & Vijg, 1993) or the *λcII* gene (Jakubczak et al., 1996; Swiger et al., 1999).

3.6 Transgenic animal models

3.6.1 LacI *transgenic model — The Big Blue® construct*

The *lacI* model is commercially available as the Big Blue® mouse and Big Blue® rat from Stratagene, La Jolla, California, USA. The *lacI* mouse model system was first described by Kohler et al. (1991a) and contains about 30–40 copies of the λLIZα shuttle vector (45.6 kb long; see Fig. 3) in a head-to-tail fashion at a single locus on chromosome 4 of the Big Blue® mouse.

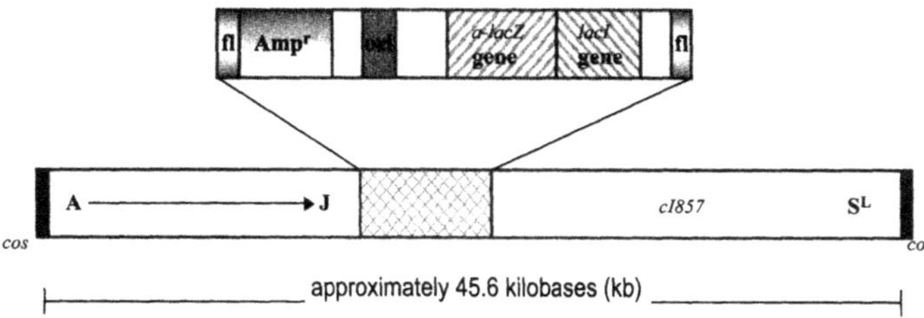

Fig. 3. Scheme of λLIZ shuttle vector *lacI*, α-*lacZ* transgene construct. Cos sites are present at either end of the construct, "Amp^r" indicates the ampicillin resistance gene, "fl" indicates the halves of the phage replication origin and "ori" indicates the bacterial origin of replication. A-J represents phage DNA encoding virus head and tail; S^L encodes for proteins that are needed for entering host cells (i.e. *E. coli*). *cI*857 represents a temperature-sensitive mutation in the *cI* gene. *cI* encodes a protein that is essential for lysogeny; with the existing mutation, the protein is not functional at 37 °C. (Adapted from Stratagene, 2002)

In the Big Blue® rat model (Dycaico et al., 1994), 15–20 copies of the λLIZα shuttle vector are present per haploid genome (Gollapudi et al., 1998).

3.6.2 lacZ *transgenic mouse model — the Muta™Mouse construct*

The Muta™Mouse was originally described by Gossen et al. (1989) and features a genomic integration of a bacteriophage λ vector (λgt10) containing the entire bacterial *lacZ* gene (Gossen et al., 1989; see Fig. 4). The vector is about 47 kb long, whereas the *lacZ* gene consists of about 3100 bp. The *lacZ* mouse model (strain 40.6) contains about 80 copies of the vector in a head-to-tail fashion (Gossen et al., 1989) at chromosome 3 (Blakey et al., 1995).

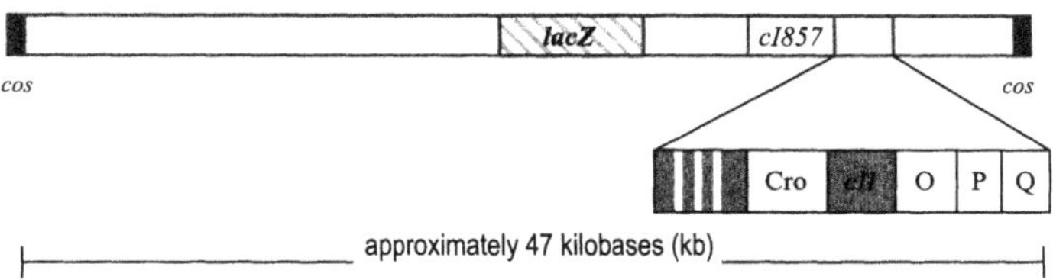

Fig. 4. Scheme of the *lacZ* construct showing the *lacZ* gene and the *λcII* gene, along with the other genes *cro*, *O*, *P* and *Q*. (Adapted from Stratagene, 2002)

4. THE TRANSGENIC MUTAGENICITY ASSAY — METHODOLOGY USING AS EXAMPLES *LACI* AND *LACZ* MUTATION MODELS

4.1 Treatment schemes

Transgenic animals containing the mutational target genes are exposed to chemical mutagens (or radiation). The appropriate treatment protocol concerning treatment times and time to tissue collection is discussed in section 6.2 and has been the subject of recent discussion at the International Workshop on Genotoxicity Testing (IWGT) (Heddle et al., 2000; Thybaud et al., 2003). One of the basic assumptions of the assay is that during treatment with a mutagen, the target gene will be damaged along with the DNA of the rodents' genomes in a proportional way.

4.2 Collection of (target) tissues and isolation of DNA

After a subsequent waiting period that allows fixation of the DNA damage into gene mutations, the animal is sacrificed and the target tissue isolated (see Fig. 5). High-molecular-weight genomic DNA has to be isolated from the target tissue according to standard procedures (e.g. Kohler et al., 1990; Vijg & Douglas, 1996; Nohmi et al., 2000). Special care has to be taken that the DNA is not damaged during this preparation. DNA fragments in the preparation should be considerably greater than 50 kb to ensure that in most instances there is no breakage between the cos sites of the vector (Vijg & Douglas, 1996).

4.3 Recovery of the DNA construct of the shuttle vector from the genomic DNA and in vitro packaging

The key development in the use of transgenic animals in gene mutation assays was the rescue of the integrated vector from the animal genome and the detection of gene mutations in vitro (Fig. 5).

In the Big Blue® model, the shuttle vector is recovered from the animal genomic DNA by mixing with an in vitro λ packaging extract

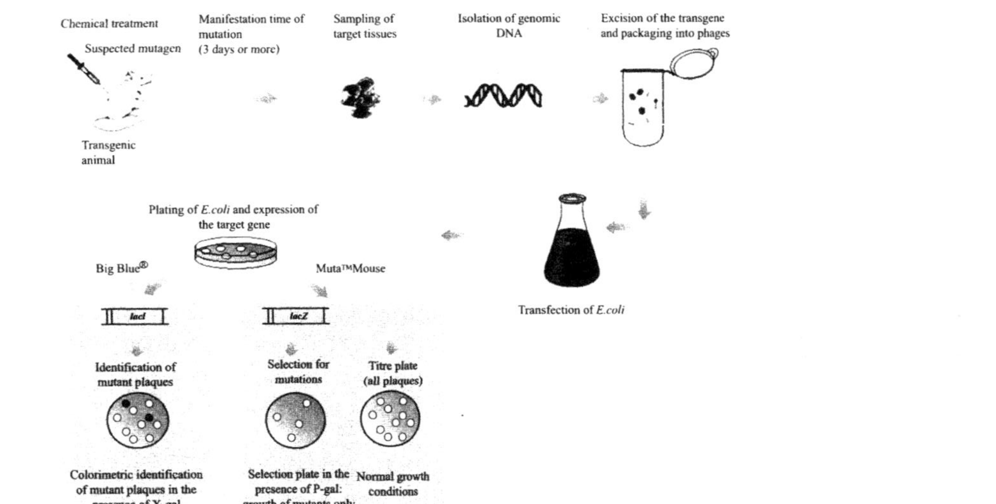

Fig. 5. General scheme of the various steps involved in the transgenic animal mutation assays. Both assays were originally performed using a colorimetric identification of mutants (see section 4.5). However, for the Muta™Mouse (*lacZ* gene) assay, this method was difficult to evaluate and has been widely replaced by the positive selection method, according to Vijg & Douglas (1996), which is shown here (see also Fig. 6). X-gal = 5-bromo-4-chloro-3-indolyl-β-D-galactopyranoside; P-gal = phenyl- β-D-galactopyranoside.

free of all known restriction systems that have been shown to reduce rescue efficiency (Kohler et al., 1991a).

Gossen et al. (1989) were the first to describe the efficient rescue of the integrated shuttle vectors from transgenic mice (Muta™Mouse). In current protocols, enzymes in the packaging extract excise the vector and insert it into a λ phage head to produce an infectious phage particle. Full-length λ DNA is packaged into individual phage particles (Vijg & Douglas, 1996).

4.4 Infection into bacteria and plating

To provide a convenient selection mechanism, *E. coli* — the bacterial host — are then infected with the phage particles (Fig. 5), which efficiently deliver the λ vectors containing the target gene into the bacterium.

The different mutation systems also need different strains of *E. coli* as hosts for the chromogenic detection of mutations: the Big Blue® system uses the *E. coli* strain SCS-8 (*lacZΔM15*), and the Muta™Mouse uses the *E. coli* strain C *ΔLacZ⁻, galE⁻ recA⁻*, pAA119 (Gossen et al., 1992; Vijg & Douglas, 1996).

4.5 Detection and quantification of mutations

Originally, the screening step is on agar plates that contain a chromogenic substrate (5-bromo-4-chloro-3-indolyl-β-D-galactopyranoside, or X-gal). The *lacZ* gene codes for the enzyme β-galactosidase. Due to the low substrate specificity, it also cleaves X-gal, which is used in the transgenic mutation detection systems (Fig. 5). More recently, the detection of mutants was improved by the use of positive selection methods.

4.5.1 *Big Blue® system*

In a non-mutant (wild type) *lacI*, gene expression occurs and results in the transcription of the LacI repressor protein. This protein binds to the operator region of the *lacZ* gene and suppresses transcription of that gene, resulting in no active β-galactosidase. The chromogenic substrate X-gal cannot be cleaved, and the plaques remain clear. On the other hand, if a mutation has occurred in the *lacI* gene, no functional repressor protein will be produced, and the *lacZ* gene α-fragment (*N*-terminus) will be expressed. The *lacZ* α-fragment combines with the carboxy terminus of the enzyme (produced by the bacterial host cell), forming the active β-galactosidase enzyme, which cleaves the chromogenic compound. The plaques are blue. Thus, mutation in the DNA in the *lacI/lacZ* gene system can be quantified as the mutant ratio or the ratio of blue plaques (mutated target genes) to colourless plaques (non-mutated target genes) (Kohler et al., 1991a).

The intensity of the blue colour, however, depends on the residual functionality of the LacI protein. Plaques that are a very light blue may therefore be missed during screening (de Boer & Glickman, 1998). The standardized assay uses four colour standards (the CM series) that consist of four mutants with increasing colour intensity (Rogers et al., 1995).

New improved protocols have been developed that improve efficiency and permit the effective measurement of mutants utilizing the Big Blue® system (Bielas, 2002). Modifications of the standard protocol concerning the medium used, the density of plated bacteria and the agarose content of the X-gal top layer resulted in a reduced plaque area but increased colour intensity.

In case of doubt, mutants can be confirmed by replating and, if necessary, by sequencing.

4.5.2 Muta™Mouse system

In a non-mutant (wild type) *lacZ* gene, gene expression occurs and results in the transcription of an active β-galactosidase.

The chromogenic substrate X-gal, which is incorporated in the agar, will be cleaved, and the plaques that eventually form become blue. On the other hand, if a mutation has occurred in the *lacZ* gene, gene expression will result in a non-functional β-galactosidase, the chromogenic substrate will not be cleaved and the plaques will remain clear. Thus, damage to the DNA in the *lacZ* gene can be quantified as the mutant ratio or the ratio of clear colourless or light blue plaques (mutant target genes) to blue plaques (non-mutated target genes) (Gossen et al., 1989).

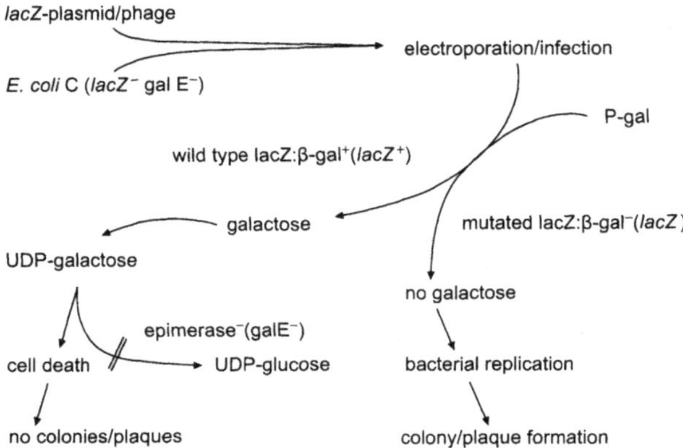

Fig. 6. Positive selection system for *lacZ⁻* phages (or plasmids). *galE⁻ lacZ⁻ E. coli* cells receiving the wild-type *lacZ* gene convert P-gal to galactose and are unable to grow, as they cannot convert the toxic UDP-galactose into the normal metabolite UDP-glucose. Only cells containing a *lacZ⁻* phage (or plasmid) will be able to form colonies and plaques, respectively, on the selective plate. Figure adapted from Vijg & Douglas (1996). β-gal = β-galactosidase; galE = uridine diphosphate galactose-4-epimerase.

Due to difficulties in the evaluation of clear plaques and the time-consuming scoring (200 000 plaques should be scored), this system has been replaced with a selection assay using phenyl-β-D-galactopyranoside (P-gal), in which only mutant particles form plaques (Gossen et al., 1992; Gossen & Vijg, 1993; Dean & Myhr, 1994). In this positive selection system, the *E. coli* strain C $\Delta LacZ^-$, *galE⁻ recA⁻* host harbours a plasmid (pAA119) overexpressing galactose kinase and transferase, but lacking the gene for β-galactosidase (*lacZ⁻*) and having a non-functional uridine diphosphate (UDP) galactose epimerase (Mientjes et al., 1996; see Fig. 6). The total number of phage-transfected bacteria (titre) is determined separately under non-selective conditions. Thus, the ratio between plaque formation under selective conditions (mutated) and plaque formation under non-selective conditions (total number: mutated + non-mutated) is a measure of mutagenicity.

5. FURTHER TRANSGENIC MUTAGENICITY ASSAYS

Although the analyses in Part II of this document are based on the Big Blue® and Muta™Mouse assays, this document does not intend to recommend *lacI* and *lacZ* of Big Blue® and Muta™Mouse as standard transgenic mutation assays. There are other systems and refinements that have been developed, in particular using positive selection, which are now available for testing. Some of these systems are mentioned below.

An international workshop of experts (IWGT) convened to discuss transgenic mutation assays in Washington, DC, USA, in 1999. It was accepted that the *lacI*, *lacZ* (phage and plasmid), *cII* and *gpt* delta assays should be considered suitable when performed under standard conditions (Heddle et al., 2000; Thybaud et al., 2003). Other systems, such as the ΦX174 transgenic model, are also briefly described here.

5.1 *λcII* assay (Big Blue® and Muta™Mouse)

The *cII* gene (294 bp) is a λ phage gene (see Fig. 4 and Fig. 7). The level of cII protein plays a central role in the lytic-lysogenic commitment made by λ phage upon infection of *E. coli.* The cII protein activates the transcription of the genes of the cI repressor protein and of the λ phage integrase protein, which are required for the establishment of lysogeny. The amount of cII protein itself is negatively regulated by the *E. coli* Hfl (high frequency of lysogeny) protease, which degrades the cII protein. If cII levels are high, λ phage will lysogenize; if levels are low, the phage will enter the lytic pathway. Upon λ phage infection of an *hfl*⁻ host, the levels of cII remain high, and all of the phage lysogenize, resulting in the absence of discernible plaques on agar plates containing *hfl*⁻ bacterial lawns. The selecting *E. coli* strains, designated G1250 or G1225, are *hfl*⁻; only when an infecting phage is *cI*⁻ or *cII*⁻ will λ phage enter the lytic pathway and form plaques. It was found that phage containing mutations specifically in the *cII* gene can be selected by plating on G1250 or G1225 at 24 °C. To determine the total number of plaques screened, a dilution of the infected host strain is incubated at 37 °C (see Fig. 8) (Stratagene, 2002).

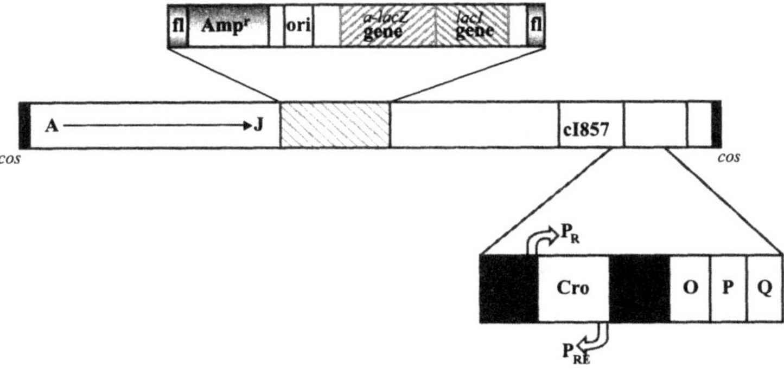

Fig. 7. The Big Blue® λLIZ shuttle vector showing the *cII* gene, along with the other genes (*cro*, *O*, *P* and *Q*) that are transcribed as a polycistronic message from the P_R promoter. The gene for the cI repressor protein contains the temperature-sensitive *cI*857 mutation that disables the cI repressor protein at 37 °C, allowing titrating in the G1250 or G1225 host strain. (Adapted from Stratagene, 2002)

The major advantage of *cII* versus colour selection systems is that they are less labour-intensive and cheaper because of the positive selection instead of colour screening for various types of mutations, including base changes and frameshifts. As *cII* is a λ gene, it can be used in the Big Blue® system (Jakubczak et al., 1996) and Muta™Mouse (Swiger et al., 1999; Swiger, 2001) but not in *gpt* delta (see below), because the *cII* gene is inactivated by the insertion of the chi sequence. It is particularly useful for Big Blue® models for which no positive selection models exist for the *lacI* gene. In addition, the *cII* gene can easily be sequenced to identify the mutational spectra. For example, *lacZ* is in excess of 3.0 kb, whereas the *cII* gene, at 294 bp, can be sequenced cost-effectively. The availability of two reporter genes (*lacZ* and *cII*, or *lacI* and *cII*) in the same DNA sample provides a method of detecting "jackpot" mutations without sequencing (Swiger et al., 1999; Heddle et al., 2000). Recently, the *cII* assay has been the assay of choice in both Muta™Mouse and Big Blue® mice or rats (see Appendix 2).

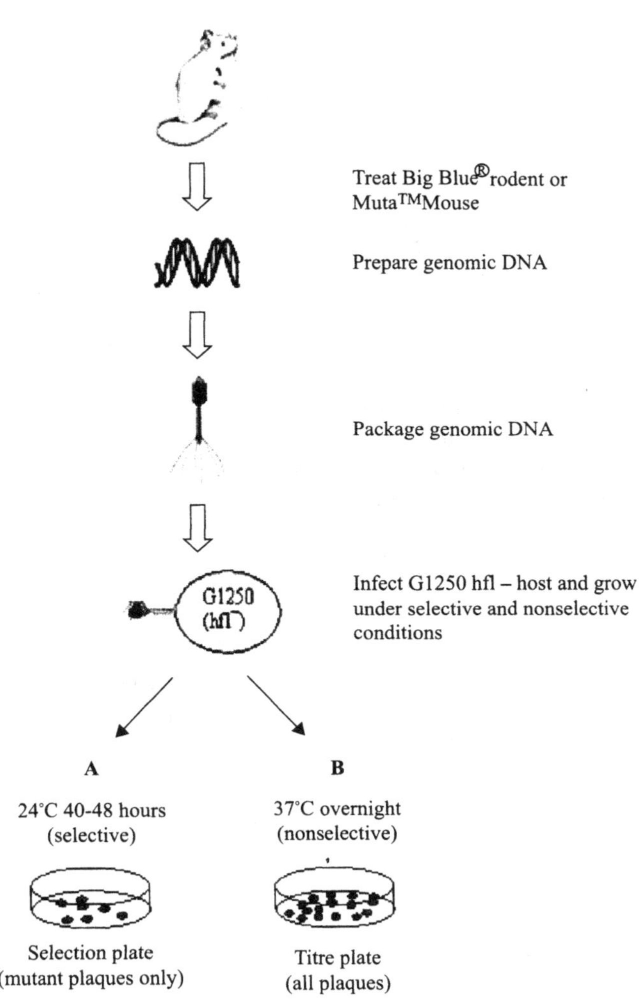

Fig. 8. *cII* selection methodology. (Adapted from Stratagene, 2002)

5.2 *gpt* delta model

5.2.1 gpt *delta rodents*

The *gpt* delta transgenic mouse model developed by Nohmi et al. (1996; see Fig. 9) features the incorporation of two different positive selection models in the transgene: the *gpt* gene of *E. coli* for point mutations (base substitutions and frameshifts) or short deletions (6-thioguanine [6-TG] selection) and Spi$^-$ selection for larger deletions of 1–10 000 bp (Horiguchi et al., 1999; Nohmi et al., 2000; Nohmi & Masumura, 2004, 2005). The *gpt* delta mice were obtained after microinjection of λEG10 phage DNA (48 kb; Figs. 9 and 10) into C57BL/6J oocytes obtained after superovulation. They carry about 80 copies of the transgene in a head-to-tail fashion at a single site of chromosome 17. The *gpt* delta mice are maintained as homozygotes and carry about 160 copies of λEG10 DNA per diploid (Masumura et al., 1999a). The coding region of *gpt* is 456 bp, which is convenient for the rapid identification of gene mutations by sequencing. Further, the positive selection system is more convenient than conventional colour selection. The *gpt* delta rat was established with the same transgene (i.e. λEG10), based on Sprague-Dawley rats (Hayashi et al., 2003), and carries 10 copies per haploid genome in chromosome 4q24-q31. The transgenic rat is maintained as a heterozygote because homozygotes are not viable.

5.2.2 *6-TG selection*

In the *gpt* delta models, the 6-TG selection method was used for detection of point mutations and small deletions. Since the product of the wild-type *gpt* converts 6-TG to a toxic substance, only cells with an inactive *gpt* gene product can survive on a plate containing 6-TG. Thus, *E. coli gpt* mutant cells can be positively selected using 6-TG.

5.2.3 *Spi$^-$ selection*

To efficiently detect deletion mutations, Spi$^-$ selection (**s**ensitive to **P2** **i**nterference) has been introduced in the transgenic animal mutation assays (Nohmi et al., 1996, 1999, 2000; Nohmi & Masumura, 2004). This selection is unique in that it preferentially and positively selects deletion mutants of λ phage (Ikeda et al.,

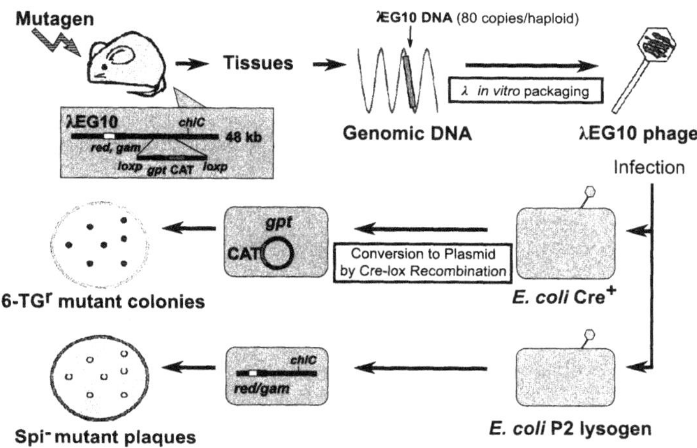

Fig. 9. Protocols of *gpt* delta transgenic mouse mutagenicity assay. Two distinct *E. coli* host cells are infected with the rescued λEG10 phages: one is *E. coli* strain YG6020 expressing Cre recombinase for 6-TG selection, and the other is P2 lysogen for Spi⁻ selection. In the cells expressing Cre recombinase, λEG10 DNA is converted to plasmid carrying *gpt* and chloramphenicol acetyltransferase (*cat*) genes. The *E. coli* cells harbouring the plasmids carrying mutant *gpt* can be positively selected as bacterial colonies on the plates containing chloramphenicol and 6-TG. Mutant λEG10 phages lacking *red/gam* gene functions can be positively selected as Spi⁻ plaques in P2 lysogens. Using *gpt* and Spi⁻ selections, the frequencies of point mutations and deletions in vivo can be compared in the same DNA samples. (Reprinted from Nohmi et al., 2000, with permission from Elsevier)

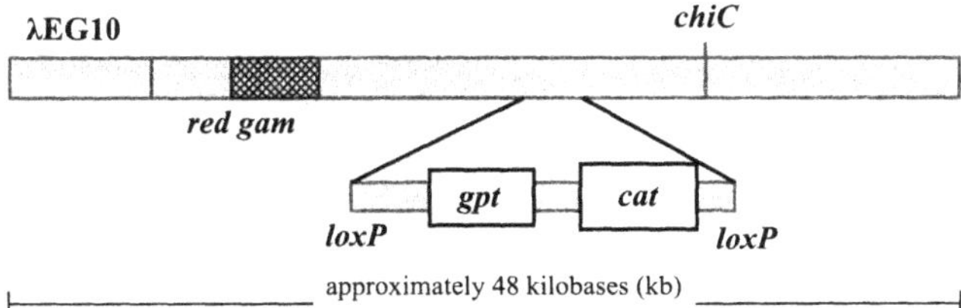

Fig. 10. Scheme of the λEG10 construct for the *gpt* delta rodent assay. (Adapted from Nohmi & Masumura, 2005, with permission from Elsevier)

1995). The selection is insensitive with respect to spontaneous or induced base change mutations and hence highlights the low incidence of deletions. Because of the size limitation for *in vitro* packaging reactions (it must have two cos sites separated by 38–51 kb of DNA), deletions detected by Spi^- selection are thought to be usually less than 10 kb. With all of these assays, some deletions starting in one copy of the vector and ending in another are theoretically recoverable but are indistinguishable from deletions arising at the corresponding sites within a single copy. Because the sites at which this could occur are small in relation to the size of the vector, it is thought that such deletions are rarely recovered and contribute little to the mutant frequency. However, the 80 copies of λEG10 DNA (each about 48 kb) still hold potential targets of about 3.8 megabases (Mb) (Nohmi et al., 2000; Nohmi & Masumura, 2004).

5.3 The *lacZ* plasmid model

The *lacZ* plasmid mouse differs from the bacteriophage-based transgenic mutagenicity systems in that a plasmid shuttle vector, not a viral vector, is recovered from genomic DNA. The *lacZ* plasmid mouse, also known as pUR288 (C57BL/6-Tg(lacZpl)60Vij/J; stock number 002754), carries the *lacZ* gene of *E. coli* as a reporter gene in the pUR288 shuttle plasmid vector in C57BL/6 mice (Gossen et al., 1995; Dollé et al., 1996). Approximately 20 copies of the pUR288 plasmid have been integrated head to tail on chromosome 11 in "line 30," whereas "line 60" harbours plasmids at chromosomes 3 and 4 (Vijg et al., 1997). The plasmid is about 5 kb long, whereas the *lacZ* reporter gene is 3100 bp.

The reason that plasmids were not initially used as the vector of choice for transgenic models was the notoriously low transformation efficiencies obtained with plasmids excised from their integrated state in the mammalian genome. Using magnetic beads coupled to the LacI repressor protein, which selectively bind to the operator sequence in front of the *lacZ* gene to recover the plasmids from genomic DNA, solved this problem (Gossen et al., 1993). The (P-gal) positive selection method is identical to that used for the bacteriophage *lacZ* model.

The background mutation frequency of this plasmid model lies in the same range as that reported for the other *lacZ* models. The characteristics of plasmid rescue allow the detection of a broad range of mutational events, including point mutations, internal deletions, insertions and chromosomal rearrangements with a breakpoint in the *lacZ* target gene and one elsewhere in the mouse genome (Dollé & Vijg, 2002). All mutational events are recovered as long as the origin of replication, the operator sequence and the ampicillin resistance gene are not affected.

The plasmid model has several advantages. Its high rescue efficiency allows the determination of mutation frequencies in small cell or tissue samples. Finally, and most importantly, the recovery of plasmids does not require a set size bracketed by required genes, permitting the detection of large deletions that are not detectable in other transgenic assays.

5.4 Upcoming transgenic models for which no extensive data are available

5.4.1 *ΦX174 transgenic mouse model*

The ΦX174 model uses the bacteriophage ΦX174*am3cs70* genome as a recovery vector in C57BL/6J transgenic mice (Burkhart et al., 1993); an embryonic cell line has also been derived from this mouse (Chen et al., 1999). Since the genome is only 5.4 kb in length (Sanger et al., 1978), the vector is recovered by electroporation, after restriction enzyme digestion and circularization by ligation. Recently, a forward assay has been developed for gene *A*, which has 40 identified target sites within the 500 *N*-terminal bases of gene *A*. All six base pair substitutions are detected as missense mutations (Valentine et al., 2002, 2004); both the original reversion assay in gene *E* and this forward assay are selective assays.

Features of this model include the historic use of ΦX174 as a genetic system (RIVM, 2000), freely available animals, an inexpensive vector recovery mechanism and a unique method for identifying mutations fixed in vivo that improves sensitivity by discarding mutations fixed in recovery bacteria.

In vivo mutations are identified by "single burst analysis" (SBA) (Delongchamp et al., 2001; Malling & Delongchamp, 2001; Malling et al., 2003; Valentine et al., 2004). SBA determines the number of progeny phage from a single electroporated bacterium by distributing aliquots of electroporated bacteria before phage growth is allowed; plating each aliquot separately identifies large numbers of plaques with the same mutation (a single burst of viral progeny from a single cell) if the mutation was fixed in the DNA before electroporation. Recovered mutant plaques not meeting cut-off criteria for the number of plaques per aliquot are discarded as having been fixed in vitro. A direct comparison of this assay using SBA with the *lacI* transgene is in progress (C.R. Valentine, personal communication, 2004).

Disadvantages of the ΦX174 assays (either forward [ΦX174*A*] or reverse [ΦX174*E*]) are that only sense base pair substitutions are detected and that SBA in its present form is highly laborious, requiring 96 agar plates for each sample analysed. An attempt to convert the assay to a 96-well format was only moderately successful (Slattery & Valentine, 2003).

5.4.2 λsupF transgenic mouse

The λsupF transgenic mouse (Leach et al., 1996a, 1996b) carries 80–100 copies of a λ phage vector carrying the tRNA nonsense codon suppressor gene, *supF*. The reporter gene is smaller than any of the other reporter genes used, having a coding region of only 85 bp; this enables a rapid identification of the gene mutations by sequencing.

5.4.3 pKZ1 transgenic recombination model

The pKZ1 transgenic recombination model provides an assay to study somatic intrachromosomal recombination (SICR) as a mutation end-point. SICR is associated with non-homologous end-joining repair of double-strand breaks and can result in chromosomal inversions and deletions, both of which are common chromosomal aberrations identified in cancers. The pKZ1 transgenic mouse was originally described by Matsuoka et al. (1991) and has been developed as a mutagenesis model by Sykes et al. (1998, 1999, 2001) and Hooker et al. (2002, 2004a, 2004b). Briefly, pKZ1 mice have the *E. coli* β-galactosidase gene (*lacZ*) in inverse orientation to a chicken β-actin enhancer/promoter complex incorporated into each cell. Histochemical detection of the *E. coli* β-galactosidase protein in frozen tissue sections can occur if SICR inverts the *lacZ* gene into the correct transcriptional orientation with respect to the enhancer/ promoter complex. A direct comparison can be made for the mutagenic effect of cyclophosphamide in spleen for Big Blue® and pKZ1 mice, as the same protocol of chemical administration and subsequent analysis was used. An induction of inversions was observed in pKZ1 spleen at doses that were 4 orders of magnitude lower (Sykes et al., 1998) than doses that had previously been shown to induce point mutations in the Big Blue® mutagenesis assay (Kohler et al., 1991b). Non-linear dose–responses have been observed in pKZ1 spleen for low doses of X-radiation (Hooker et al., 2004a) and etoposide (Hooker et al., 2002). For some of the low doses studied, there was a significant reduction in inversions below endogenous frequency. By contrast with most mutation assays, the pKZ1 assay has a very high endogenous inversion frequency in spleen tissue of approximately 1.5×10^{-4}. This high frequency makes it feasible to observe a reduction below endogenous frequency.

5.4.4 rpsL *transgenic mouse model*

A transgenic mouse model has been developed using C57BL/6J mice and the *rpsL* gene in a shuttle plasmid (Gondo et al., 1996). An *E. coli* shuttle plasmid pML4 carrying the kanamycine-resistant gene next to the *rpsL* (*strA*) gene was used. As the reporter gene is only 375 bp long, the *rpsL* transgenic mouse model is very suitable for monitoring mutation spectra. Further, a positive selection system is used (streptomycin selection), which is more convenient than the conventional colour screening (RIVM, 2000).

6. PARAMETERS AND CRITERIA FOR VALID EXPERIMENTAL DESIGN

6.1 Determinants studied using transgenic mutation assays

6.1.1 *Types of mutations*

Mutations detectable by the *lacI* (Big Blue®) and *lacZ* (Muta™Mouse) transgenes are point mutations, deletions and insertions less than 8 kb. For the Big Blue® model, the mutations will be predominantly in the *lacI* gene, but they may also occur in the *lacI* promoter region or in the *lacZ* operator. Since in vitro packaging requires that the λ vector has a minimum size of approximately 37 kb, deletions or insertions larger than about 8 kb are not detectable in these systems. Furthermore, insertions larger than about 4 kb would not be detectable because of packaging restrictions. Therefore, clastogenic agents, which predominantly induce large deletions, yield low responses in the bacteriophage *lacI* and *lacZ* assays. The plasmid-based transgenic mouse can detect such deletions (Vijg & Douglas, 1996), as can the *gpt* delta rodents following Spi$^-$ selection (Nohmi et al., 2000; Nohmi & Masumura, 2004, 2005).

6.1.2 *Overall mutant/mutation frequency*

The *mutant frequency* is the ratio of the number of mutant plaques to the total number of plaques, corrected for the dilution factor. The mutant frequency is not necessarily the best reflection of the rate of mutation events, since a single mutation event during development or cell growth can produce a large pool of cells carrying the same mutation (clonal expansion), which would give a high mutant frequency ("jackpot" mutation) that did not reflect the true frequency of mutation events. *Mutation frequency* is defined as the mutant frequency corrected for clonal expansion. In transgenic mutagenesis assays, mutants are scored by counting plaque numbers (or colonies); mutations can be identified by sequencing the target gene. Clonal expansion can be estimated by correcting for mutations that recur one or more times in a given tissue from a given mouse, by counting a given mutation only once per animal per tissue.

However, this correction is based on the assumption that multiple identical mutations in the same tissue of the same animal arise from a single mutation.

6.1.3 Spontaneous mutant frequency

To determine increases in mutant frequencies, knowledge about the background (spontaneous) mutant frequency is important (i.e. the mutant frequency observed in untreated animals). The *lacZ*, *lacI* and *cII* models all exhibit similar high levels of spontaneous mutant frequencies, in the 10^{-5} range in most tissues (Vijg & Douglas, 1996; de Boer et al., 1998), which is approximately 5- to 8-fold higher than for the endogenous human *HPRT* gene (Cole & Skopek, 1994). However, the spontaneous mutant frequency in the *gpt* delta mouse is 5–10 × 10^{-6} (Nohmi & Masumura, 2004). Thus, the spontaneous mutant frequency appears to be dependent on the reporter genes used (see also section 11.1.2).

6.1.4 Sequence analysis

In addition to the determination of mutant frequencies, the exact sequence alteration can be determined after isolating the mutant plaques and amplifying the DNA sequence of the phage particles using polymerase chain reaction. The DNA sequencing of mutations has facilitated the validation of the transgenic animal assays by comparison of mutation spectra of transgenes with those of endogenous genes.

Mutations of the reporter gene (*lacI* or *lacZ*) can be tabulated for a particular chemical or agent to obtain a spectrum. The characterization of the induced mutational changes (e.g. transversions, transitions, frameshifts or deletions) may lead to an understanding of the mechanism of the chemical damage (see chapter 11). The fingerprints detected in DNA analysis are valuable in searching for or assessing a causative substance in human carcinogenesis (molecular epidemiology) and important for understanding the mechanisms underlying mutagenesis. As well, in the case of a marginal increase in the mutation frequency, a spectrum in DNA sequence analysis could determine whether the test substance is positive or negative.

The mutation spectra can be compared with the sequence spectra of the original *lacZ* gene of *E. coli* (Kalnins et al., 1983; Tsutsui et al., 1999) or *lacI* gene of *E. coli* (Farabaugh, 1978) or against other existing databases — e.g. for spontaneous mutation data from *lacZ* (Douglas et al., 1994) and *lacI* (Kohler et al., 1991a, 1991b; Cariello et al., 1998). Due to its size (3 kb), sequencing of the *lacZ* gene requires much effort; however, sequencing is faster with the much shorter *λcII* gene (0.3 kb) (Jakubczak et al., 1996).

6.2 Criteria for valid experimental design of transgenic mutation assays

6.2.1 *Selection of the most suitable model*

All available data on general toxicity of the examined substance should be considered in the selection of the species and strain (Big Blue® available in F344 rats and different mouse strains), gender and examined organ (Heddle et al., 2000). Data (if available) on genotoxic/carcinogenic sensitivity should be analysed, as well as data on target organs in long-term studies. For transgenic studies designed to investigate the carcinogenic mechanism of a particular substance, preferably the main target organs from carcinogenicity studies in the selected species should be chosen.

If no significant differences exist between the sexes or the tested substance shows no female specificity, then male animals are normally used. Further factors in the selection of a suitable model are the route of human exposure and toxicokinetic parameters (Thybaud et al., 2003). For other hazard identification, other routes may be appropriate. Other data (e.g. physicochemical properties of the test substance) and metabolic pathways in different species should also be considered in selecting the relevant model.

6.2.2 *Duration of exposure*

Potent mutagens can be detected after a single treatment or in 5-day subacute exposure (Heddle et al., 2000). However, weak mutagens may need several treatments at the maximum tolerated dose (MTD; dose that produces toxicity, above which lethality is expected) over a prolonged exposure period to yield a significant increase in the mutation frequency. For example, 2,4-diaminotoluene resulted in a 2-fold increase in the mutant frequency in the liver after a 90-day treatment with 1000 mg/kg in the diet but not after a 30-day treatment (for comparison: 100 mg/kg induced hepatocarcinogenic effects in long-term studies; Cunningham et al., 1996). The available information indicates that mutations accumulate linearly with the number of treatments given, even for periods as long as 90 days (Heddle et al., 2000). Negative results in tests involving five or fewer treatments at the MTD are regarded as inadequate for evaluation. However, no study fully evaluates the optimum number of daily exposures. Nevertheless, based on available data, in a recent publication of the IWGT (Thybaud et al., 2003), the authors suggested daily treatments for a period of 4 weeks for producing a sufficient number of mutations by weak mutagens. Treatment times longer than 8 weeks should be employed with caution because of a possible increase in the mutant frequency due to clonal expansion or genomic instability in developing preneoplastic foci or tumours. However, alternative treatment regimens may be appropriate to meet the anticipated exposure in humans (e.g. the clinical regimen of a drug is for weekly intravenous injection for 4 weeks) and should be justified in the protocol (Thybaud et al., 2003).

6.2.3 Selection of the dose

The top dose used should produce minimal to zero mortality (i.e. MTD). This dose can be determined in dose range-finding studies using non-transgenic animals of the same strain (Mirsalis et al., 1995). Heddle et al. (2000) recommended two further dose groups: one third and two thirds of the MTD. Analysis of the mutant frequency of only the two highest dose groups would be sufficient, if all three dose groups are complete. The two lower dose groups should be analysed if the number of animals in the high dose group is reduced by increased mortality, reducing the statistical power below an acceptable level (Heddle et al., 2000).

6.2.4 Post-treatment manifestation time

The sensitivity of the transgenic test system is influenced by the post-exposure observation period.

This manifestation or fixation time is required to fix the DNA damage into an irreversible mutation (Mirsalis et al., 1995). The manifestation time for in vivo fixation of the DNA adduct or damage to mutation is dependent on the proliferation rate of the particular tissue. The proliferation rate itself may be influenced by the chemical treatment. The significance of a negative result is uncertain unless the manifestation time is known (Heddle, 1999; Sun & Heddle, 1999). A short manifestation time is acceptable in rapidly dividing tissues (e.g. bone marrow or colon mucosa). However, in tissues with low mitotic rates (e.g. brain, heart), a longer manifestation time is needed to get a maximum response (1 month or more). Fifty days may be required to detect mutation in sperm cells due to the long period needed for maturation. Mutations induced in spermatogonia stem cells may need even longer times for their manifestation (Douglas et al., 1995a).

In rapidly dividing tissues like the bone marrow, longer manifestation times might decrease the mutation frequency due to the loss of differentiated cells from the tissue. For example, *lacZ* mutation response in the male Muta™Mouse after seven daily treatments with acrylamide was compared using manifestation times of 3 and 28 days (Thybaud et al., 2003). A manifestation time of 28 days resulted in mutant frequency in the bone marrow similar to the concurrent control, while a significant increase was reported at a manifestation time of 3 days. When the treatment period was extended to 28 days, the manifestation time of 3 days resulted in the highest measured mutant frequency; in addition, the longer manifestation time of 28 days showed a doubling in the mutant frequency compared with control. Overall, the authors concluded that 1) a 28-day treatment period should allow sufficient accumulation in slowly proliferating tissues (e.g. the liver), even 3 days after the final treatment, and 2) sampling 3 days after the treatment period will ensure that mutations in rapidly dividing tissues are not lost. Therefore, in tissues for which the optimal sampling time is unknown, sampling 3 days following 28 daily treatments should be used (Heddle et al., 2000, 2003; Thybaud et al., 2003). If slowly proliferating tissues (e.g. the liver) are of particular interest, then a manifestation time greater than 3 days following the 28-day treatment period may be more appropriate (Thybaud et al., 2003).

The extended treatment protocol that has been recommended is based on the neutrality of mutations in the transgenes and so should be equally valid for all of the transgenic assays discussed here, except for the *lacZ* plasmid assay. In this assay, which, unlike the other assays, is able to detect deletions large enough to extend far beyond the transgene, many of the deletion mutations are not genetically neutral, so the optimal protocol is uncertain (Boerrigter, 1999).

6.2.5 Significance of a negative result

Provided that a suitable protocol was used, that tissue exposure can be demonstrated and that the appropriate tissues were sampled, a result that does not demonstrate a significant increase in mutant frequency in any tissue compared with data from untreated controls can be regarded as negative with confidence. Under certain circumstances, when scientifically justified, evaluation of a single tissue might be sufficient to define a negative. For a chemical for which the tissues at risk are not known from pharmacokinetic or other toxicological data, measurement of bone marrow, liver and a tissue relevant to the route of administration should be regarded as sufficient to establish a negative. For example, for an inhalation study, lung would be a suitable third tissue, whereas for oral administration, the gastrointestinal tract and oral cavity would be suitable.

6.2.6 Factors to consider when comparing the performance of mutation assays

In order to make fully valid comparisons using the sensitivity of mutagenicity assays or end-points (see section 6.2.8), it is essential that each assay also be performed under optimal conditions, using similar dose ranges and dose regimens. This approach ensures that responses are at maximal levels. Most assays for genotoxic end-points require quite rigorous optimization of key variables. The most prevalent variable is the time between the end of treatment and sampling. This consideration applies to chromosomal aberration assays as well as gene mutation assays, and for both in vitro and in vivo assays. Manifestation time is influenced by cell and tissue type and can vary considerably among tissues used in the same transgenic assay, primarily due to the rate of cell division. Optimal manifestation time is also influenced by age of animals in the *Hprt* assay.

Another critical variable affecting all toxicity studies is a sufficient number of animals or replicates, which directly affects the statistical power of a test result. Comparisons of test results that do not incorporate these considerations should be interpreted with caution.

6.2.7 Positive control

Normally, concurrent positive control animals are not necessary (except in laboratories new to these test systems), but positive control DNA should be included with each plating to confirm the validity of the method (Heddle et al., 2000; Thybaud et al., 2003).

6.2.8 Sensitivity

The sensitivity of a mutation assay can be measured in different ways, but the most fundamental issue is the ability of the assay to detect an increase in mutant frequency above background. A relevant measure of sensitivity that facilitates the comparison of assay results is the minimum effective dose of a mutagen. The accuracy of this measurement depends on the background, spontaneous frequency and, of course, the protocol used. For valid comparisons, it is essential that the optimal protocol be used for each assay, including manifestation time, number of animals or replicates and the size of each sample.

The *Hprt* assay, which has a low spontaneous frequency, often shows a higher ratio of induction of mutations than transgenic mutation assays that have higher spontaneous frequencies. However, a lower spontaneous frequency compared with the transgenic assays does not necessarily mean that the *Hprt* assay is more sensitive, since the number of mutants detected and the number of copies of the locus examined can be much larger in the transgenic assays. These additional factors may improve the statistical power of comparisons involving transgenic mutation assays. In addition to the spontaneous mutant frequency, the relative sensitivity of the assays will be determined by other factors, such as the nature of the mutations detectable (which favours most assays involving endogenous loci) and limitations on the tissues that can be analysed (which is extremely limited for most assays of endogenous loci). For many assays, including the micronucleus and other assays for chromosomal aberrations, as well as the *Hprt* assay, there is an optimal time for measurement, such that missing it will produce different (less than optimal) results. Comparisons of test results that do not incorporate these considerations are unlikely to be valid.

6.2.9 Statistics

In older studies on transgenic systems, no consensus exists about statistical evaluation of the results. Which data are considered to constitute a "positive" response? In workshops on statistical analysis of mutation data from transgenic mice (Gorelick & Thompson, 1994), the minimal study designs and appropriate statistical analysis were recommended (Mirsalis et al., 1995; Piegorsch et al., 1997; Delongchamp et al., 1999). It is estimated that at least five animals and 200 000 plaques per animal are necessary to accurately detect a 2-fold increase in the mutation frequency (Mirsalis et al., 1995).

In general, the response must be statistically significant compared with appropriate concurrent controls using appropriate statistical analysis, and the fold increase required for significance is a function of the number of animals in each group, the number of plaques analysed per animal and the degree of animal-to-animal variation observed. Further, in a properly designed study with multiple doses, a trend analysis should also be performed to assess the likely significance of the pattern of observed responses (Carr & Gorelick, 1994, 1995, 1996). However, statistical significance should not be the only determining factor for a positive response; biological relevance of the results should be considered first.

6.2.10 Analysis of DNA sequence

While not considered mandatory for mutant detection and conclusion of clearly positive and negative studies, sequence data give useful additional information. Changes in the frequency of specific classes of mutations can be a more sensitive measure of mutation induction than the overall frequency of mutations (Heddle et al., 2000). Furthermore, sequencing can be used to rule out experimental artefacts like "jackpots" or clonal events by identifying unique mutants from the same tissue (Thybaud et al., 2003; see also section 11.3 on sequencing).

6.2.11 Use of transgenic assays in the detection of gene mutations in germ cells

Mutagenic effects on male germ cells can be detected if the temporal progression of spermatogenesis and the possible delay in the timing due to toxicity are adequately considered. If spermatozoa are sampled from the epididymis, all individual stages of spermatogenesis can be evaluated by sampling at a range of times following treatment. At manifestation times longer than 50 days, mutations arising from spermatogonial stem cells can be detected; shorter manifestation times will detect mutations at progressively later stages of spermatogenesis. Higher mutant frequencies have been observed in spermatozoa at later times following treatment, indicating their origin in spermatogonial stem cells (Douglas et al., 1995a; Ashby et al., 1997). However, with agents such as *N*-ethyl-*N*nitrosourea and isopropyl methanesulfonate, mutations have also been observed at shorter times following treatment, consistent with a post-meiotic origin (Douglas et al., 1995a). Accordingly, spermatozoa should be sampled from the epididymis over a range of manifestation times that reflects all stages of spermatogenesis.

Germ cells can also be sampled from seminiferous tubules to yield a mixed population of cells covering a wider range of developmental stages, providing the opportunity to use fewer manifestation times to cover the range of developmental stages (Douglas et al., 1995a).

PART II:

COMPARISON OF THE *LACI* MODEL AND THE *LACZ* MODEL WITH CONVENTIONAL TEST SYSTEMS

7. INTRODUCTION TO PART II

7.1 Aim of the comparison and criteria for the selection of data

Since their introduction, transgenic mutation assays have been applied to a large number of compounds (Gorelick, 1995; Schmezer & Eckert, 1999; Thybaud et al., 2003). Therefore, it was decided in this EHC to compare the outcome of these assays with the outcome of conventional in vivo mutation assays (see Table 1 in chapter 2) in order to assess how they could contribute to the overall assessment of genotoxicity. Furthermore, it was investigated whether transgenic mutagenicity assays identify target organs in carcinogenicity studies.

Although there are several transgenic animal assays for testing mutagenicity, in Part II the documentation has been limited to the *lacZ* gene in the Muta™Mouse and the *lacI* gene in the Big Blue® mouse or rat, because these are the only two systems with enough data to allow comparisons to be made and data analysis to be performed.

Chapter 8 compares data on Muta™Mouse and Big Blue® to see if they give the same results for test chemicals.

Chapter 9 compares Big Blue® and Muta™Mouse (using transgenic reporter genes) with other assays currently used in genotoxicity testing, using data from exposure to different chemicals:

- a chromosomal aberration assay, the mouse bone marrow micronucleus test (OECD Test Guideline 474, an example of a widely used assay in toxicity testing; OECD, 1997a) (section 9.1);
- in vivo gene mutation assays using endogenous reporter genes — i.e. the mouse spot test (OECD Test Guideline 484; OECD, 1986a) (section 9.2.1) and *Hprt* and *Dlb-1* models (section 9.2.2);
- unscheduled DNA synthesis (UDS) assay (section 9.3); and
- genotoxicity assays in vitro (section 9.4).

Chapter 10 discusses transgenic mutation assays and carcinogenicity:

- mutagenicity in target organs as shown by Big Blue® and Muta™Mouse transgenic assays compared with carcinogenicity reported in target organs after exposure to different chemicals (section 10.1); and
- the use and acceptance of transgenic mutation assays in their ability to predict carcinogenicity (section 10.2).

Sources of the detailed information on transgenic animal mutation assays were exclusively the primary literature. Data on the mouse spot test and the mouse micronucleus assay were collected from primary literature or from reviews with sufficient details on methods and results for evaluation and comparison with other studies on the same end-point.

All other data on genotoxicity, in vitro and in vivo, as well as data on carcinogenicity in mice and rats, were extracted from secondary literature. This literature has different degrees of reliability, depending on the quality of the prepared document or the data bank. The used sources were, for example, documents prepared by IARC, the German MAK Commission and WHO/IPCS (EHCs or CICADs). Data banks like HSDB, CCRIS, IRIS and GENE-TOX (for definitions of acronyms, see abbreviation list) were also used if none of the above assessment documents was available for a particular substance or to update the available information if the documents were prepared before the year 2000.

7.2 Choice and limitations of data

Data on the Muta™Mouse or the Big Blue® mouse or rat are available for approximately 100 substances (Gorelick, 1995; Schmezer & Eckert, 1999; Heddle et al., 2000; Nohmi et al., 2000; RIVM, 2000; Thybaud et al., 2003). From these 100 substances, a selection was made here of 55 substances for which carcinogenicity studies were available. These are tabulated in this document (see Master Table, Appendix 1).

Starting points for the selection of substances from the data pool were as follows. For the comparison of target organs in transgenic animal mutation assays with target organs in long-term carcinogenicity studies in the same species (chapter 10), those substances were chosen for which transgenic animal studies were carried out on more than one target organ and data on carcinogenicity were available for the corresponding species. A few substances were then added for the investigation into non-genotoxic carcinogens and non-carcinogens.

For the comparison of results of the transgenic mouse mutation assay with results of the mouse spot test (section 9.2.1), all available data on the mouse spot test (search in secondary literature and data bank TOXLINE) were compared with data from Big Blue® mouse and Muta™Mouse, and substances were selected with data in both data pools. A similar procedure was performed for the comparison of results of the transgenic mouse mutation assay with results of the mouse bone marrow micronucleus test.

The comparison of results of the transgenic mouse/rat mutation assay with results of in vitro genotoxicity test systems (section 9.4) was performed with the available data collected for the other studies given above.

As a consequence of these criteria, studies on, for example, X-rays and radiation were not considered in this part of the document.

7.3 Validity of data on transgenic animal mutation assays

The various studies using transgenic animal mutation assays have used varying protocols that were not always as robust as the protocol recently recommended by the IWGT (Thybaud et al., 2003). This makes a comparison of the results difficult. Therefore, a pragmatic approach was used. No examination of the validity was performed if any result obtained in at least one target organ of at least one study on transgenic animals was positive for the particular test substance. However, if all available data on transgenic animals gave negative results for a substance, the validity was checked (see Table 18 in chapter 10) using the validity criteria described in section 6.2. For the analysis, it is considered that a negative result using a robust protocol should be accepted as valid.

7.4 Criteria for predictivity of transgenic assays

In order to understand the reliability of any new mutagenicity test, emphasis must first be placed on determining whether this assay produces results that are comparable with the results of existing similar mutagenicity assays. Accordingly, it is the *accuracy* of a test that is the primary consideration in terms of the detection of mutagenicity. In contrast, the ability of a mutagenicity test to detect potential carcinogenic activity is described in terms of *predictivity*, rather than accuracy.

The available data from transgenic mutation assays suggest that they accurately detect mutations, a fact confirmed by DNA sequencing of many mutant phenotypes and indicated further by the fact that many mutagens induce specific mutation spectra. Conversely, non-mutagens in other mutation assays do not induce mutations in transgenic assays, and mutants sequenced from control animals exhibit a consistent characteristic spectrum. Accordingly, the veracity of the results of well conducted transgenic animal mutation assays, in terms of the detection of gene mutagens, is very high.

In terms of carcinogenicity, the positive predictivity of transgenic animal mutation assays is as good as, or better than, that of established mutagenicity assays; however, the negative predictivity, determined from the analysis of a very small number of non-carcinogens, is low, as is the case with other mutagenicity tests.

Positive responses, by their very nature, are more readily accepted if they were obtained using protocols that were suboptimal. In contrast, negative results obtained using such a suboptimal protocol must be interpreted with caution. It should be realized that many studies using transgenic mutation assays cited in this review were not performed using the subacute exposure protocol developed recently (Thybaud et al., 2003). Accordingly, older protocols using single or low numbers of dose applications may not have been adequate to detect weak mutagenic effects.

8. COMPARISON OF THE MUTA™MOUSE AND BIG BLUE® ASSAYS

Data on both the Muta™Mouse and Big Blue® mouse or rat assays are available for 13 of 55 substances listed in this document. All of these 13 substances showed neoplastic effects in carcinogenicity studies on mice. Details of transgenic studies and carcinogenic studies are presented in the Master Table (see Appendix 1), which also includes the corresponding references.

With 9 of 13 chemicals, the mutant frequency was increased in at least one target organ in the Muta™Mouse *and* in the Big Blue® mouse, independent of the route of exposure: 2-acetylaminofluorene, benzo[*a*]pyrene, 1,3-butadiene, cyclophosphamide, 7,12-dimethylbenz[*a*]anthracene, *N*-ethyl-*N*-nitrosourea, *N*-methyl-*N*-nitrosourea, *N*-nitrosodimethylamine and urethane. In the case of 1,3-butadiene and 7,12-dimethylbenz[*a*]anthracene, the same experimental parameters (e.g. mode of administration, total dose, administration time, sampling time) have been used in both systems. Positive results were obtained with 7,12-dimethylbenz[*a*]anthracene in the skin with both systems following topical application. Discordant results were observed with 1,3-butadiene following inhalation studies using identical protocols, with positive results in bone marrow of Big Blue® mice but negative results in this organ with Muta™Mouse. However, positive results were obtained in Muta™Mouse lung following the same inhalation protocol. *N*-Ethyl-*N*-nitrosourea is generally used as a positive control in transgenic animal mutagenicity studies, and positive results have been obtained in both systems in a variety of organs following generally similar experiments.

Discordant results were obtained in liver of Muta™Mouse and Big Blue® mouse following treatment with methyl methanesulfonate. However, the treatment protocols differed in these experiments, a fact that limits the comparison.

Phenobarbital has been examined in both Muta™Mouse and Big Blue® mouse and rat. Initial results suggested that this compound was negative in Muta™Mouse but increased the mutant frequency in both the *lacI* gene and *cII* gene of the Big Blue® mouse.

Different experimental protocols were used with Muta™Mouse and Big Blue®. Subsequent sequencing of *lacI* and *cII* mutants facilitated clonal correction and yielded a negative result in the Big Blue® transgenic assay, consistent with that observed in Muta™Mouse.

In conclusion, the limited data available suggest that there is significant agreement with respect to the results obtained with the Muta™Mouse and the Big Blue® mouse or rat assay. Any observed differences between the Muta™Mouse and the Big Blue® mouse are likely to be attributable to the different experimental designs used in the particular studies, rather than to differences in the sensitivity of the transgenic reporter genes per se.

9. TRANSGENIC ASSAYS — COMPARISON WITH OTHER ASSAYS

9.1 The Muta™Mouse assay and the Big Blue® mouse assay versus the mouse bone marrow micronucleus test[1]

The mouse bone marrow micronucleus test is one of several available in vivo mammalian test systems for the detection of structural and numerical chromosomal aberrations (Heddle et al., 1983; Mavournin et al., 1990; Shelby et al., 1993; Morita et al., 1997a, 1997b). Documentation of the test procedure and guidance for evaluating the results are given in the OECD Test Guideline 474 (OECD, 1997a). This test is routinely used with a widespread acceptance by industry and authorities.

Since both point mutations and chromosomal aberrations (micronuclei) may be induced by a single agent, some overlap of results is to be expected. Nevertheless, the advantage of the transgenic assays is that they are not limited to bone marrow, as is the standard micronucleus assay.

A comparison of the transgenic mouse assays with the mouse bone marrow micronucleus test highlights the fact that different genotoxic end-points are studied in these two systems. In transgenic mouse assays, point mutations and small insertions and deletions are detected, whereas in the mouse bone marrow assay, chromosome breakage leads to light microscopically visible micronuclei resulting from chromosome fragments or from whole chromosomes.

Results from the mouse bone marrow micronucleus test were compared with results from the Big Blue® mouse and the Muta™Mouse assays for 44 substances (see Table 2).

---

[1] A previous version of this section has been published (Wahnschaffe et al., 2005a).

Table 2. Comparison of results of the transgenic mouse assay and mouse bone marrow micronucleus test[a]

Substance	Results in carcinogenicity studies on mice [IARC evaluation][b]	Results of transgenic assays [in all studied organs] only in bone marrow		Mouse bone marrow micronucleus test			Further assays detecting chromosomal aberration	
		Muta™-Mouse	Big Blue® mouse	Results	Agreement with Muta™-Mouse[c]	Agreement with Big Blue® mouse[c]	in vitro[d]	in vivo
2-Acetylamino-fluorene	Positive [no evaluation]	[+] nd	[+] nd	+	Yes	Yes	+	+ (micronuclei, rat)
4-Acetylamino-fluorene	nd [no evaluation]	[+] nd	nd	±	Inconclusive	na	nd	nd
Acrylamide	Positive [2A]	[+] +	nd	±	Inconclusive	na	++	+ (micronuclei, mouse germ cells) + (cytogenetic, mouse) + (dominant lethal, mouse & rat) + (heritable translo-cation, mouse)
Aflatoxin B1	Positive [1]	nd	[+] nd	+	na	Yes	+	+ (micronuclei, rat) + (cytogenetic, rat & mouse) ± (dominant lethal, mouse)

Table 2 (Contd)

Substance	Results in carcinogenicity studies on mice [IARC evaluation][b]	Results of transgenic assays [in all studied organs] only in bone marrow		Mouse bone marrow micronucleus test			Further assays detecting chromosomal aberration	
		Muta™-Mouse	Big Blue® mouse	Results	Agreement with Muta™-Mouse[c]	Agreement with Big Blue® mouse[c]	in vitro[d]	in vivo
(contd)								+ (dominant lethal, rat)
4-Amino-biphenyl	Positive [1]	[+] +	nd	+	Yes	na	nd	− (micronuclei, rat)
2-Amino-3,4-dimethyl-imidazo[4,5-*f*]quinoxaline (MeIQx)	Positive [2B]	nd	[+] nd	−	na	No	±	+ (cytogenetic, rat)
2-Amino-3-methyl-imidazo[4,5-*f*]quinoline (IQ)	Positive [2A]	[+] nd	nd	−	No	na	±	− (cytogenetic, mouse) + (cytogenetic, rat)
2-Amino-1-methyl-6-phenylimidazo-[4,5-*b*]pyridine (PhIP)	Positive [2B]	[+] nd	nd	±	Inconclusive	na	+	+ (cytogenetic, mouse)

Table 2 (Contd)

Substance	Results in carcinogenicity studies on mice [IARC evaluation][b]	Results of transgenic assays [in all studied organs] only in bone marrow		Mouse bone marrow micronucleus test			Further assays detecting chromosomal aberration	
		Muta™-Mouse	Big Blue® mouse	Results	Agreement with Muta™-Mouse[c]	Agreement with Big Blue® mouse[c]	in vitro[d]	in vivo
ortho-Anisidine	Positive [2B]	nd	[+] nd	–	na	No	+	– (micronuclei, rat)
Asbestos crocidolite	Positive [1]	nd	[+] nd	–	na	No	++	nd
Benzene	Positive [1]	nd	[+] +	+	na	Yes	±	+ (micronuclei, rat) + (cytogenetic, rat, mouse & human) – (cytogenetic, *Drosophila*)
Benzo[*a*]py-rene	Positive [2A]	[+] +	[+] nd	+	Yes	Yes	++	+ (micronuclei, rat) + (cytogenetic, mouse & hamster) – (cytogenetic, rat) + (dominant lethal, mouse)

Table 2 (Contd)

Substance	Results in carcinogenicity studies on mice [IARC evaluation][b]	Results of transgenic assays [in all studied organs] only in bone marrow		Mouse bone marrow micronucleus test			Further assays detecting chromosomal aberration	
		Muta™-Mouse	Big Blue® mouse	Results	Agreement with Muta™-Mouse[c]	Agreement with Big Blue® mouse[c]	in vitro[d]	in vivo
Bromomethane	Negative [3]	[−] −	nd	+	No	na	±	+ (micronuclei, rat) ± (micronuclei, human) − (cytogenetic, rat) − (dominant lethal, rat)
1,3-Butadiene	Positive [2A]	[+] −	[+] +	+	Yes	Yes	nd	− (micronuclei, rat & human) + (cytogenetic, human & mouse) + (dominant lethal, mouse) − (dominant lethal, rat) + (heritable translocation, mouse)
Chlorambucil	Positive [1]	[+] +	nd	+	Yes	na	+	+ (micronuclei, rat) + (cytogenetic, rat) ± (cytogenetic, human)

Table 2 (Contd)

Substance	Results in carcinogenicity studies on mice [IARC evaluation][b]	Results of transgenic assays [in all studied organs] only in bone marrow		Mouse bone marrow micronucleus test			Further assays detecting chromosomal aberration	
		Muta™-Mouse	Big Blue® mouse	Results	Agreement with Muta™-Mouse[c]	Agreement with Big Blue® mouse[c]	in vitro[d]	in vivo
Chloroform	Positive but non-genotoxic [2B]	nd	[−] nd	−	na	Yes	−	+ (micronuclei, rat) + (cytogenetic, rat) ± (cytogenetic, mouse)
Cyclophos-phamide	Positive [1]	[+] +	[+] −	+	Yes	Yes	++	+ (micronuclei, rat) + (cytogenetic, rat, mouse & human) + (dominant lethal, mouse & rat) + (heritable translocation, *Drosophila*)
2,4-Diamino-toluene	Positive [2B]	nd	[+] nd	−	na	No	±	+ (micronuclei, rat) − (dominant lethal, mouse)
2,6-Diamino-toluene	Negative (valid study) [no evaluation]	nd	[−] nd	+	na	No	++	− (cytogenetic, rodents)

Table 2 (Contd)

Substance	Results in carcinogenicity studies on mice [IARC evaluation][b]	Results of transgenic assays [in all studied organs] only in bone marrow		Mouse bone marrow micronucleus test			Further assays detecting chromosomal aberration	
		Muta™-Mouse	Big Blue® mouse	Results	Agreement with Muta™-Mouse[c]	Agreement with Big Blue® mouse[c]	in vitro[d]	in vivo
1,2-Dibromo-ethane	Positive [2A]	[+] nd	nd	–	No	na	++	– (cytogenetic, mouse) – (dominant lethal, rat & mouse)
1,2-Dibromo-3-chloropropane	Positive [2B]	[(+)] nd	nd	+	Yes	na	+	+ (micronuclei, rat) + (cytogenetic, rat) + (dominant lethal, rat) – (dominant lethal, mouse) ± (heritable trans-location, *Drosophila*)
1,2-Dichloro-ethane	Positive [2B]	[–] nd	nd	–	Yes	na	++	– (dominant lethal, mouse)
Di-(2-ethyl-hexyl) phthal-ate	Positive but non-genotoxic [2B]	nd	[–] nd	–	na	Yes	––	– (cytogenetic, rat) + (cytogenetic, hamster) ± (dominant lethal, mouse)

Table 2 (Contd)

Substance	Results in carcinogenicity studies on mice [IARC evaluation][b]	Results of transgenic assays [in all studied organs] only in bone marrow		Mouse bone marrow micronucleus test			Further assays detecting chromosomal aberration	
		Muta™-Mouse	Big Blue® mouse	Results	Agreement with Muta™-Mouse[c]	Agreement with Big Blue® mouse[c]	in vitro[d]	in vivo
7,12-Dimethyl-benz[*a*]anthra-cene	Positive [no evaluation]	[+] +	[+] +	+	Yes	Yes	+	+ (micronuclei, rat) + (cytogenetic, rodent)
Ethylene oxide	Positive [1]	nd	[+] +	+	na	Yes	++	+ (micronuclei, rat & human) + (cytogenetic, rat, mouse & human) + (dominant lethal, rat & mouse) + (heritable translo-cation, mouse & *Drosophila*)

Table 2 (Contd)

Substance	Results in carcinogenicity studies on mice [IARC evaluation][b]	Results of transgenic assays [in all studied organs] only in bone marrow		Mouse bone marrow micronucleus test			Further assays detecting chromosomal aberration	
		Muta™-Mouse	Big Blue® mouse	Results	Agreement with Muta™-Mouse[c]	Agreement with Big Blue® mouse[c]	in vitro[d]	in vivo
Ethyl methane-sulfonate	Positive [2B]	[+] +	nd	+	Yes	na	++	+ (micronuclei, rat) + (cytogenetic, mouse) + (dominant lethal, rat & mouse) + (heritable translo-cation, mouse & *Drosophila*)
N-Ethyl-*N*-nitrosourea (ENU)	Positive [2A]	[+] +	[+] nd	+	Yes	Yes	++	+ (cytogenetic, rat & mouse) + (heritable translo-cation, *Drosophila*)
Hydrazine & hydrazine sulfate	Positive [2B]	[−] −	nd	+	No	na	±	− (cytogenetic, mouse) + (cytogenetic, *Drosophila*) − (dominant lethal, mouse)

Table 2 (Contd)

Substance	Results in carcinogenicity studies on mice [IARC evaluation][b]	Results of transgenic assays [in all studied organs] only in bone marrow		Mouse bone marrow micronucleus test			Further assays detecting chromosomal aberration	
		Muta™-Mouse	Big Blue® mouse	Results	Agreement with Muta™-Mouse[c]	Agreement with Big Blue® mouse[c]	in vitro[d]	in vivo
Methyl methane-sulfonate	Positive [2B]	[(+)] –	[–] nd	+	Yes	No	++	+ (micronuclei, rat) + (cytogenetic, mouse) + (dominant lethal, mouse) + (heritable translocation, mouse)
N-Methyl-*N*'-nitro-*N*-nitroso-guanidine (MNNG)	Positive [2A]	[+] –	nd	+	Yes	na	++	+ (micronuclei, rat) + (cytogenetic, mouse) – (dominant lethal, mouse)
N-Methyl-*N*-nitrosourea	Positive [2A]	[+] nd	[+] nd	+	Yes	Yes	++	+ (cytogenetic, *Drosophila*) + (dominant lethal, mouse) + (heritable translo-cation, *Drosophila*)

Table 2 (Contd)

Substance	Results in carcinogenicity studies on mice [IARC evaluation][b]	Results of transgenic assays [in all studied organs] only in bone marrow		Mouse bone marrow micronucleus test			Further assays detecting chromosomal aberration	
		Muta™-Mouse	Big Blue® mouse	Results	Agreement with Muta™-Mouse[c]	Agreement with Big Blue® mouse[c]	in vitro[d]	in vivo
Mitomycin C	Positive [2B]	[−] −	nd	+	No	na	++	+ (micronuclei, rat) + (cytogenetic, mouse) + (dominant lethal, rodents) + (heritable translo-cation, mouse)
4-Nitroquino-line 1-oxide	Positive [no evaluation]	[+] +	nd	+	Yes	na	++	+ (micronuclei, rat)
N-Nitrosodi-ethylamine	Positive [2A]	[+] −	nd	−	No	na	+	− (micronuclei, rat) − (dominant lethal, mouse) ± (heritable translo-cation, *Drosophila*)

Table 2 (Contd)

Substance	Results in carcinogenicity studies on mice [IARC evaluation][b]	Results of transgenic assays [in all studied organs] only in bone marrow		Mouse bone marrow micronucleus test			Further assays detecting chromosomal aberration	
		Muta™-Mouse	Big Blue® mouse	Results	Agreement with Muta™-Mouse[c]	Agreement with Big Blue® mouse[c]	in vitro[d]	in vivo
N-Nitrosodi-methylamine	Positive [2A]	[+] nd	[+] –	+	Yes	Yes	+	± (micronuclei, rat) – (cytogenetic, mammals) ± (dominant lethal, rodents) + (heritable translo-cation, *Drosophila*)
N-Nitrosodi-*n*-propylamine	Positive [2B]	[+] +	nd	–	No	na	+	nd
Phenobarbital	Positive [2B]	[–] nd	[(+)] nd	±	Inconclusive	Inconclusive	±	– (cytogenetic, mouse)

Table 2 (Contd)

Substance	Results in carcinogenicity studies on mice [IARC evaluation][b]	Results of transgenic assays [in all studied organs] only in bone marrow		Mouse bone marrow micronucleus test			Further assays detecting chromosomal aberration	
		Muta™-Mouse	Big Blue® mouse	Results	Agreement with Muta™-Mouse[c]	Agreement with Big Blue® mouse[c]	in vitro[d]	in vivo
Procarbazine	Positive [2A]	[+] +	nd	+	Yes	na	–	+ (cytogenetic, mouse) ± (dominant lethal, mouse) + (dominant lethal, *Drosophila*) – (heritable translo-cation, mouse & *Drosophila*)
β-Propiolac-tone	Positive [2B]	[+] –	nd	–	No	na	+	+ (cytogenetic, plant) + (heritable translo-cation, *Drosophila*)
Quinoline	Positive [2A]	[+] –	nd	+	Yes	na	+	– (micronuclei, rat) – (cytogenetic, mouse) + (cytogenetic, rat)
Tetrachloro-methane	Positive but non-genotoxic [2B]	[–] nd	nd	–	Yes	na	±	– (cytogenetic, rat & mouse)

Table 2 (Contd)

Substance	Results in carcinogenicity studies on mice [IARC evaluation][b]	Results of transgenic assays [in all studied organs] only in bone marrow		Mouse bone marrow micronucleus test			Further assays detecting chromosomal aberration	
		Muta™-Mouse	Big Blue® mouse	Results	Agreement with Muta™-Mouse[c]	Agreement with Big Blue® mouse[c]	in vitro[d]	in vivo
Trichloro-ethylene	Positive [2A]	[−] −	nd	+	No	na	±	± (micronuclei, rat) − (cytogenetic, rat & mouse) − (dominant lethal, mouse)
Tris(2,3-di-bromopropyl)-phosphate	Positive [2A]	nd	[+] nd	−	na	No	±	+ (micronuclei, hamster) − (cytogenetic, rat & mouse) + (heritable translo-cation, *Drosophila*)
Urethane	Positive [2B]	[+] +	[+] nd	+	Yes	Yes	+	+ (micronuclei, rat) ± (cytogenetic, *Drosophila*) + (heritable translo-cation, *Drosophila*)

Table 2 (Contd)

−: negative study results; +: positive (for transgenic mouse assays, at least one examined organ shows an increased mutant/mutation frequency); ++: majority of results are positive concerning two or more end-points in in vitro studies; (+): study result weakly positive; ±: inconclusive result; nd: no data available; na: not applicable because one test not done; −−: majority of results are negative concerning two or more end-points

[a] All data in this table are taken from the Master Table in Appendix 1.

[b] IARC categories: Group 1, the agent is carcinogenic to humans; Group 2A, the agent is probably carcinogenic to humans; Group 2B, the agent is possibly carcinogenic to humans; Group 3, the agent is not classifiable as to its carcinogenicity to humans; Group 4, the agent is probably not carcinogenic to humans.

[c] All organs.

[d] Chromosomal aberration assays (cytogenetic assay in mammalian cells; micronucleus assay in mammalian cells; cytogenetic assay and/or testing of aneuploidy in fungi, e.g. *Saccharomyces cerevisiae*).

9.1.1 The mouse bone marrow micronucleus test: principles and procedures

Micronuclei are chromatin-containing bodies in the cytoplasm arising from acentric chromosome fragments or from whole chromosomes that were not incorporated into the daughter nuclei during the last stages of mitosis. The presence of chromosome fragments is associated with the clastogenic (chromosome-breaking) activity of the test substance, whereas the presence of a whole chromosome is indicative of aneuploidy. The difference in size of the micronucleus could therefore be an indicator for clastogenicity (generally small micronucleus) or aneugenicity (usually large micronucleus). However, the size of the micronucleus is an imprecise measure. Micronuclei can be distinguished by further criteria — for example, by identification of the presence of a kinetochore or centromeric DNA, indicating aneugenic activity. Overall, an increase in micronuclei is a measure of induced structural or numerical chromosomal aberrations.

In the last three decades, toxicologists have routinely used the mouse bone marrow micronucleus test because 1) it is part of the regulatory toxicology in the submission or approval procedure for chemicals and drugs and 2) it has advantages in speed, simplicity and cost-effectiveness in comparison with the other in vivo systems for testing chromosomal aberrations — for example, the chromosomal aberration assay (OECD Test Guideline 475: OECD, 1997b).

9.1.2 Comparison of data from the mouse bone marrow micronucleus test and transgenic mouse test

9.1.2.1 Bone marrow

Bearing in mind that the mouse bone marrow micronucleus test detects clastogenic effects while the transgenic assays detect primarily gene mutations, it is interesting to assess how the two test systems respond in the same target organ (bone marrow). For most (13 of 25 with data on the bone marrow) of the substances (4-aminobiphenyl, benzene, benzo[*a*]pyrene, 1,3-butadiene, chlorambucil, cyclophosphamide, 7,12-dimethylbenz[*a*]anthracene, ethyl methanesulfonate, ethylene oxide, *N*-ethyl-*N*-nitrosourea, 4-nitroquinoline 1-oxide, procarbazine, urethane), positive results have been obtained, both in the transgenic assays and in the micronucleus test, indicating that these substances cause mutagenic *and* clastogenic effects. Only three substances (1,3-butadiene, cyclophosphamide and 7,12-dimethylbenz[*a*]anthracene) were tested in both transgenic assays, but only 7,12-dimethylbenz[*a*]anthracene gave a positive result in Muta™Mouse and Big Blue® mouse assays (Table 2). The differences between the results of the other two substances with these assays might be related to experimental design.

For *N*-nitrosodiethylamine and β-propiolactone, negative results have been obtained in bone marrow for both micronucleus and transgenic test systems, although these two compounds are carcinogens. These compounds are positive in other organs in transgenic animal mutagenicity assays, which suggests that the reactive entities do not reach the bone marrow.

There is one carcinogenic substance, *N*-nitrosodi-*n*-propylamine, with a negative micronucleus test and a positive result in the bone marrow of the Muta™Mouse, which might indicate gene mutagenic rather than clastogenic activity in this organ.

There are, however, several substances for which positive results have been obtained in the bone marrow micronucleus test but negative results in the bone marrow transgenic assay. Examples are bromomethane, hydrazine and hydrazine sulfate, methyl methanesulfonate, *N*-methyl-*N*'-nitro-*N*-nitrosoguanidine, mitomycin C, *N*-nitrosodimethylamine, quinoline and trichloroethylene. Some of the negative results in the transgenic assays may be explained by the test conditions. For example, the negative Muta™Mouse assay on mitomycin C was not conducted at the MTD or with repeated administration, whereas with a more rigorous design it was detected with *gpt* delta and Spi$^-$ (Takeiri et al., 2003). Therefore, especially for weak mutagens, this might be a reason for negative results in the transgenic assays. On the other hand, a substance like mitomycin C might induce more chromosomal aberrations than gene mutations in bone marrow and therefore gives a negative result in the transgene but increased incidence in micronuclei of the same Muta™Mouse (Suzuki et al., 1993).

In practice, for applying the transgenic assays, the negative results in the bone marrow may be of minor relevance, because there are other target organs that may be more sensitive than the bone marrow. For example, methyl methanesulfonate, *N*-methyl-*N'*-nitro-*N*-nitrosoguanidine, *N*-nitrosodimethylamine and quinoline were negative in the transgenic assays in the bone marrow but positive in other organs. Nevertheless, one must be aware of problems in sensitivity of the transgenic test system, which may be due to testing conditions or to the restriction of the test system for detecting only small deletions.

9.1.2.2 *All organs*

One aspect to be considered in comparing both test systems is that the micronucleus test is restricted to one target organ, which may not be reached by unstable reactive compounds or reactive metabolites. In contrast, in the transgenic assays, any target organ may be investigated. Therefore, if one looks at all target organs, the transgenic assay may have a higher sensitivity. This is indeed the case. For quite a number of carcinogenic substances, positive results have been obtained in at least one of the transgenic tests, whereas the mouse bone marrow micronucleus test was negative. Examples are 2-amino-3,4-dimethylimidazo[4,5-*f*]quinoxaline, 2-amino-3-methylimidazo[4,5-*f*]quinoline, *o*-anisidine, asbestos crocidolite, 2,4-diaminotoluene, 1,2-dibromoethane, *N*-nitrosodiethylamine, *N*-nitrosodi-*n*-propylamine, β-propiolactone and tris(2,3-dibromopropyl)phosphate. Local effects that appear at the site of application of the test substance can be detected by the transgenic mutagenicity assays, but not by the mouse bone marrow micronucleus test. Examples are the alkylating substance β-propiolactone or asbestos crocidolite (see Master Table, Appendix 1).

There are only 2 of 44 substances without carcinogenic effects in mice in Table 2: bromomethane and 2,6-diaminotoluene. Both gave positive results in the micronucleus test, but no mutagenic activity was detected in the transgenic mouse assays (for analysed organs, see Master Table in Appendix 1).

9.1.3 ***Predictivity of the transgenic animal mutagenicity assays and the mouse bone marrow micronucleus test for carcinogenicity***

The sensitivity, specificity and predictive values for carcinogenicity of the Muta™Mouse assay and the Big Blue® mouse assay combined and the mouse bone marrow micronucleus test are documented in Table 3. In the present study, data on 43 substances were available concerning carcinogenicity in mice *and* mutagenic effects in transgenic mice as well as mutagenic effects in the mouse bone marrow micronucleus test (Table 2). The 3 (of 43) substances with inconclusive results in the mouse bone marrow micronucleus assay (phenobarbital, acrylamide and 2-amino-1-methyl-6-phenylimidazo-[4,5-*b*]pyridine) were not included in the final calculation and in the comparison of the micronucleus test with the transgenic mouse assay.

Table 3. Characteristics of the Muta™Mouse assay and the Big Blue® mouse assay for predicting mouse carcinogenicity in comparison with the micronucleus test[a]

Term[b]	Calculation for the mouse bone marrow micronucleus test	Calculation for Muta™Mouse and/or Big Blue® mouse combined[c]
Sensitivity	63% (24/38)	82% (31/38)
Specificity	0 (0/2)	100% (2/2)
Positive predictivity	92% (24/26)	100% (31/31)
Negative predictivity	0 (0/14)	22% (2/9)
Overall accuracy	60% (24/40)	83% (33/40)

[a] Carcinogens with genotoxic and non-genotoxic mechanisms were considered, as well as non-carcinogenic substances; only data on mice were used.

[b] Sensitivity = % of carcinogens with a positive result in the specified test system (STS); specificity = % of non-carcinogens with a negative result in the STS; positive predictivity = % of positive results in the STS that are carcinogens; negative predictivity = % of negative results in the STS that are non-carcinogens; overall accuracy = % of chemicals tested where STS results agree with carcinogenicity results.

[c] Judged as positive in transgenic assays if positive in at least one of the two test systems. Weak positive results in transgenic mouse assays were judged as positive.

Although the data pool mentioned in this document is not sufficient for a comprehensive comparison (low number of examples, especially for specificity and negative predictivity), some differences were apparent between the two test systems. The overall accuracy of the micronucleus test is lower than that of the transgenic mouse assays. This is mainly due to 14 negative results in the micronucleus test system (negative in the micronucleus test but positive in carcinogenicity studies), influencing the terms sensitivity and negative predictivity. Three of these negative results in the micronucleus test are obtained with carcinogenic substances (chloroform, di-(2-

ethylhexyl) phthalate and tetrachloromethane) for which carcinogenic effects are considered to be via a non-genotoxic (non-DNA-reactive) mechanism. However, chloroform, di-(2-ethylhexyl) phthalate and tetrachloromethane gave negative results in transgenic mice, so the comparison of both test systems is not essentially affected, and the evaluation as "non-genotoxic" is supported. For the other 11 substances with negative results in the micronucleus test, these results are readily explainable: *o*-anisidine (mutagenic/carcinogenic effects are restricted to the bladder), 2-amino-3,4-dimethylimidazo[4,5-*f*]quinoxaline and 2-amino-3-methylimidazo[4,5-*f*]quinoline (bone marrow is presumably not the target organ of genotoxicity in mice, and substance is more gene mutagenic than clastogenic), asbestos (local genotoxic/carcinogenic effects in the lung), 2,4-diaminotoluene (target organ liver, presumably not bone marrow), 1,2-dibromoethane (more local than systemic effects), *N*-nitrosodiethylamine (target organ liver, more gene mutagenic than clastogenic), *N*-nitrosodi-*n*-propylamine (presumably more gene mutagenic than clastogenic), β-propiolactone (mainly local effects and less systemic effects in bone marrow) and tris(2,3-dibromopropyl)-phosphate (systemic effects not related to bone marrow).

Since the database of Table 2 contains only two non-carcinogenic compounds, only sensitivity and positive predictivity are reliable. The other parameters are interesting, but have little statistical value.

Negative predictivity is also low in the transgenic mouse assay due to false-negative results for six carcinogenic substances; for four of them — 1,2-dichloroethane, hydrazine, mitomycin C and trichloroethylene — genotoxic mechanisms are presumed. For hydrazine (no repeated application) and 1,2-dichloroethane and trichloroethylene (MTD not reached), limitations on the experimental design might be the reason for the negative results. Mitomycin C is clearly more clastogenic than gene mutagenic. However, the *lacI/lacZ* transgenic assay was not conducted at the MTD or with repeated administration. It is worth noting that the *gpt* delta transgenic mouse with repeated dose detected deletions larger than 2 kb induced by mitomycin C (Takeiri et al., 2003).

For three of the nine substances with negative results in the transgenic mouse assay, the carcinogenic effects in mice were attributed to non-genotoxic mechanisms: chloroform, di-(2-ethylhexyl) phthalate and tetrachloromethane (see also section 10.2). All gave negative results in transgenic mice; however, the protocols were optimized only for chloroform.

Only two substances with negative results in long-term carcinogenicity studies are available in the data pool of Table 2: bromomethane and 2,6-diaminotoluene. Both gave correct negative results in the transgenic mouse assay (although of limited validity) but false-positive results in the micronucleus test (see term specificity in Table 3 and footnote b for explanation).

The differences between the two test systems might be due to the fact that 1) unequal genotoxic end-points are investigated (chromosome mutation in the micronucleus test versus gene mutation in the transgenic mouse assay), 2) organotrophy of genotoxic effects (especially when bone marrow is not the target organ of mutagenicity) might play an essential role and 3) transgenic animal mutagenicity assay conditions in the different systems may not be optimal for mutation detection. Consequently, these two assays are complementary in their value, since they measure different aspects of genotoxicity. Table 4 shows that sensitivity for both tests together reaches 89%, exceeding the 82% for transgenic animal mutagenicity assays alone or the 63% for the micronucleus assay alone (Table 3). The further testing of non-carcinogenic compounds will clarify the value of using the two tests together, since that will allow a measure of specificity of the combined tests.

9.1.4 Comparison of both test systems

a) Sensitivity of the test system

In comparison with other test systems in genotoxicity testing using endogenous target structures, the spontaneous mutant frequency in the transgenic mouse assay is relatively high. This might be related to the fact that the transgene is bacterial DNA (high methylation rate) or that the transgene is silent and that no transcription-related repair occurs as for endogenous genes, which are more efficiently repaired (RIVM, 2000). In the mouse bone marrow micronucleus test, the spontaneous rate of micronuclei is low, ranging between 1 and 3 polychromatic erythrocytes (PCEs) with micronuclei per 1000 PCEs. The frequency of chromosomal aberrations is not directly comparable with a mutant frequency.

Table 4. Comparison of the results in the mouse bone marrow micronucleus assay and transgenic mouse assays for carcinogens[a]

	Positive results in the mouse bone marrow micronucleus assay	Negative results in the mouse bone marrow micronucleus assay
Positive results in the Muta™Mouse and/or the Big Blue® mouse assay	21 (55%)	10 (26%)
Negative results in the Muta™Mouse and/or the Big Blue® mouse assay	3 (8%)	4 (11%)

[a] Data taken from Table 2; only substances (n = 38) with positive results in carcinogenicity studies on mice were used; weak positive results in transgenic mouse assays were judged as positive; three substances were not included in this table (although positive results in carcinogenicity are given) because of inconclusive results in the mouse bone marrow micronucleus test.

Comparing the target organs and cells at risk at the time of exposure, the mouse micronucleus test is restricted to one target organ, the bone marrow, especially to the erythroblasts. This limitation is not present for transgenic mouse assays: target cells are cells in all organs (Nohmi et al., 2000).

b) Considerations of animal welfare

Both test systems are similar in the number of animals used for a valid test. The minimal number of mice needed in the mouse bone marrow assay is 25 per gender (three dose levels, vehicle control, positive control; five mice per group) using a treatment schedule with two or more applications at 24-h intervals and sampling 1824 h following the final treatment (or one application and two sampling times). In the limit test (for a test substance demonstrating no toxicity), only one dose level of 2000 mg/kg of body weight is necessary (OECD, 1997a).

In transgenic mutation assays, 20 animals (three dose groups and one concurrent vehicle control group in laboratories that have already established this test system) are recommended per species and gender (Mirsalis et al., 1995; Heddle et al., 2000). In terms of animal welfare, it is also desirable that more than one in vivo genotoxicity assay, such as the transgenic mouse assay and micronucleus assay, be merged, using the same animals for both assays. It is possible to use transgenic mice/rats for long-term carcinogenicity bioassays as well.

c) Cost-effectiveness

Due to the simplicity of the mouse bone marrow micronucleus assay and the use of systems for automated analysis, this test is less expensive than the transgenic mouse assay.

A comparison of both test systems is presented in Table 5.

9.1.5 Conclusions

The differences between the two test systems might be due to the fact that 1) the transgenic animal mutation assay, which is not yet routinely used in toxicological screening, is not equivalent to the micronucleus test, because different genetic end-points are examined (chromosome mutation versus gene mutation), 2) the transgenic animal mutagenicity assay has advantages over the micronucleus test, in that it is not restricted to one target organ and detects local as well as systemic mutagenic effects, and 3) transgenic animal mutagenicity assay conditions may not be optimal for mutation detection (solved with the recommended protocol of Thybaud et al., 2003). However, these two assays are complementary in their value, since they measure different aspects of genotoxicity, and both systems were found to have a place in mutagenicity testing and to complement each other.

Table 5. Comparison of the mouse bone marrow micronucleus assay with transgenic mouse models (Muta™Mouse and the Big Blue® assay)[a]

	Mouse bone marrow micronucleus test (1,2)	Transgenic mouse mutation assay (3,4)
Type of end-point	Detects light microscopically visible micronuclei resulting from whole chromosomes or chromosome fragments following chromosome breakage	Detects gene mutation and small deletions or insertions
Regulatory use	Widespread acceptance (OECD guideline established since 1983)	Not routinely used by the industry in toxicological screening; OECD guideline proposed (5)
Background mutation rate	Spontaneous incidence of micronuclei is low (about 0.3%) and almost uniform	High spontaneous rate of mutations compared with other mutation assays
Negative predictivity	Low negative predictivity for cancer (Table 3, but limited database)	Low negative predictivity for cancer (Table 3, but limited database)
Implementation	Simplicity of the test system; easily recognized end-point	Higher complexity of the test system (target cells in mice and expression of mutagenic effects in bacteria; vector system needed)
Toxicokinetics and metabolism	Restrictions in toxicokinetics: unstable test substance or the toxic metabolites may not reach the bone marrow, the only target organ	No restrictions after absorption and distribution of the test substance
Target tissue	Restricted to erythroblasts in the bone marrow	No tissue restriction
Dependency of effects on application route	Only systemic effects can be detected	Local as well as systemic mutagenic effects can be detected
Number of animals	Five animals per gender per dose recommended	Five animals per gender per dose recommended
Restrictions on the model used	Some recommendations are given in OECD Test Guideline 474; no limitation concerning species, strain, gender, age of animals, exposure duration	Limitations: Muta™Mouse assay only one species and one strain; Big Blue® two species (mouse and rat) but one (rat) or two strains (mouse); no limitations on other parameters

Table 5 (Contd)

	Mouse bone marrow micronucleus test (1,2)	Transgenic mouse mutation assay (3,4)
Costs	Less expensive due to the simplicity of the test system	More expensive test system
Molecular mechanism	Mechanisms of the induction of micronuclei originating from chromosome fragments could not be resolved; only fragment and whole chromosome can be distinguished	Detection of the "molecular signature" of a particular mutagenic substance by DNA sequence analysis with standardized methods
Parallel examination of different genetic end-points	Combination with other genotoxic end-points is not recommended but possible if results of the micronucleus test are not influenced and vice versa	The transgenic mouse assay can be combined with other in vivo genotoxic end-points in the same animal (micronuclei, chromosomal aberration, UDS) if results in the transgenic assays are not influenced and vice versa
Type of mutational target	In situ end-point	Target genes are integrated parts of foreign DNA and consequently no "normal" mutational target, no expression

[a] References are as follows: 1) Heddle et al., 1983; 2) Mavournin et al., 1990; 3) RIVM, 2000; 4) Nohmi et al., 2000; and 5) Health Canada, 2004.

9.2 The Muta™Mouse assay and the Big Blue® mouse or rat assay versus assays using endogenous reporter genes

9.2.1 Results in the mouse spot test compared with those from transgenic animals[2]

In the mid-1980s, the mouse spot test (Fahrig, 1977) was suggested as a complementary in vivo test to the bacterial mutagenicity assay for detection of gene mutagenic substances and as a confirmatory test for the identification of carcinogens (Styles & Penman, 1985). The mouse spot test, an in vivo assay, has been used

[2] A previous version of this section has been published (Wahnschaffe et al., 2005b).

to assess a number of chemicals (see, for example, Table 6). It is at present the only in vivo mammalian test system capable of detecting somatic mutations according to OECD guidelines (OECD Test Guideline 484: OECD, 1986a). However, to achieve an acceptable sensitivity, a large number of animals are necessary, and it is therefore an expensive type of test and seldom used.

Here, the results of in vivo testing of a number of chemicals using the mouse spot test are compared with results from Big Blue® mouse (*lacI*) and Muta™Mouse (*lacZ*).

9.2.1.1 *Description of the mouse spot test*

In the spot test, mouse embryos that are heterozygous for different recessive coat colour genes are treated in utero on gestation days 9–11 with the test substance. The exposed embryo at gestation day 10 contains about 150–200 melanoblasts, and each melanoblast has four coat colour genes under study (Fahrig, 1977; Russell et al., 1981). The in utero exposure may result in an alteration or loss of a specific wild-type allele in a pigment precursor cell, resulting in a colour spot in the coat of the adult animal. The frequency of spots is compared with the frequency in sham-exposed controls (Fahrig, 1977; OECD, 1986a).

In the mouse spot test, there are four possible mechanisms that can lead to the expression of recessive coat colour alleles: 1) gene mutation in the wild-type allele, 2) deficiency (large or small) of a chromosomal segment involving the wild-type allele, 3) non-disjunctional or other loss of the chromosome carrying the wild-type allele and 4) somatic recombination causing the marker to become homozygous (Russell et al., 1981). Thus, both gene mutation and clastogenic effects are detected by this test system.

9.2.1.2 *Comparison of the mouse spot test with transgenic mouse model systems*

A literature search was made for chemicals that had been tested using the spot test *and* the Muta™Mouse assay (n = 22) or the Big Blue® mouse assay (n = 9) or both transgenic mutation assays (n = 8). The results are given in Table 6.

Table 6. Comparison of results of the transgenic mouse assays and mouse spot test[a]

Substance	Results in carcinogenicity studies on mice [IARC evaluation][b]	Results in transgenic assays		Results of mouse spot test	Agreement of mouse spot test with		Further gene mutation assays	
		Muta™-Mouse	Big Blue® mouse		Muta™-Mouse	Big Blue® mouse	in vitro[c]	in vivo
2-Acetylamino-fluorene	Positive [no evaluation]	+	+	+	Yes	Yes	++	± (*Drosophila*, SLRL)
4-Acetylamino-fluorene	nd [no evaluation]	+	nd	−	No	na	++	nd
Acrylamide	Positive [2A]	+	nd	+	Yes	na	++	+ (specific locus, mouse) + (host mediated) + (*Drosophila*, SLRL) + (*Drosophila*, somat.)
2-Amino-3-methyl-imidazo[4,5-*f*]quinoline (IQ)	Positive [2A]	+	nd	−	No	na	++	+ (*Hprt*, rat) + (host mediated) + (*Drosophila*, SLRL) + (*Drosophila*, somat.)
Benzo[*a*]py-rene	Positive [2A]	+	+	+	Yes	Yes	++	+ (*Drosophila*, somat.) − (*Drosophila*, SLRL)
1,3-Butadiene	Positive [2A]	+	+	+	Yes	Yes	−	+ (*HPRT*, human & *Hprt*, mouse) + (*Drosophila*, somat.)

Table 6 (Contd)

Substance	Results in carcinogenicity studies on mice [IARC evaluation][b]	Results in transgenic assays		Results of mouse spot test	Agreement of mouse spot test with		Further gene mutation assays	
		Muta™-Mouse	Big Blue® mouse		Muta™-Mouse	Big Blue® mouse	in vitro[c]	in vivo
(contd)								− (*Drosophila*, SLRL)
Cyclophos-phamide	Positive [1]	+	+	+	Yes	Yes	++	+ (host mediated) + (*Drosophila*, somat.) + (*Drosophila*, SLRL)
1,2-Dibromo-3-chloropropane	Positive [2B]	(+)	nd	+	(Yes)	na	++	− (specific locus test, mouse) + (*Drosophila*, somat.) + (*Drosophila*, SLRL)
1,2-Dichloro-ethane	Positive [2B]	−	nd	±	Inconclusive	na	++	− (host mediated) + (*Drosophila*, somat.) + (*Drosophila*, SLRL)
Di-(2-ethyl-hexyl) phthalate	Positive but non-genotoxic	nd	−	−	na	Yes	−−	± (*Drosophila*, somat.) − (*Drosophila*, SLRL)
Ethyl methane-sulfonate	Positive [2B]	+	nd	+	Yes	na	++	+ (specific locus test, mouse) + (host mediated) + (*Drosophila*, SLRL)

Table 6 (Contd)

Substance	Results in carcinogenicity studies on mice [IARC evaluation][b]	Results in transgenic assays		Results of mouse spot test	Agreement of mouse spot test with		Further gene mutation assays	
		Muta™-Mouse	Big Blue® mouse		Muta™-Mouse	Big Blue® mouse	in vitro[c]	in vivo
N-Ethyl-*N*-nitrosourea	Positive [2A]	+	+	+	Yes	Yes	++	+ (specific locus test) + (*Hprt*, mouse) + (*Drosophila*, SLRL)
Hydrazine & hydrazine sulfate	Positive [2B]	–	nd	+	No	na	++	+ (host mediated) + (*Drosophila*, SLRL) + (*Drosophila*, somat.)
Methyl methane-sulfonate	Positive [2B]	(+)	–	+	Yes	No	++	± (specific locus test) + (*Hprt*, rat) + (host mediated) + (*Drosophila*, somat.) + (*Drosophila*, SLRL)
N-Methyl-*N'*-nitro-*N*-nitroso-guanidine	Positive [2A]	+	nd	+	Yes	na	++	+ (host mediated) + (*Drosophila*, somat.) + (*Drosophila*, SLRL)
N-Methyl-*N*-nitrosourea	Positive [2A]	+	+	+	Yes	Yes	++	+ (host mediated) + (*Drosophila*, SLRL)
Mitomycin C	Positive [2B]	–	nd	+	No	na	++	+ (specific locus test) + (host mediated)

Table 6 (Contd)

Substance	Results in carcinogenicity studies on mice [IARC evaluation][b]	Results in transgenic assays		Results of mouse spot test	Agreement of mouse spot test with		Further gene mutation assays	
		Muta™-Mouse	Big Blue® mouse		Muta™-Mouse	Big Blue® mouse	in vitro[c]	in vivo
(contd)								+ (*Drosophila*, SLRL)
4-Nitroquinoline 1-oxide	Positive [no evaluation]	+	nd	+	Yes	na	++	+ (host mediated)
N-Nitrosodiethylamine	Positive [2A]	+	nd	+	Yes	na	++	− (specific locus test) + (host mediated) + (*Drosophila*, SLRL)
N-Nitrosodimethylamine	Positive [2A]	+	+	+	Yes	Yes	++	+ (*Drosophila*, SLRL)
Procarbazine	Positive [2A]	+	nd	+	Yes	na	++	+ (specific locus test) + (host mediated) + (*Drosophila*, somat.) + (*Drosophila*, SLRL)
N-Propyl-*N*-nitrosourea	nd [no evaluation]	+	nd	+	Yes	na	++	nd
Trichloroethylene	Positive [2A]	−	nd	±	Inconclusive	na	(+)	− (host mediated)

Table 6 (Contd)

–: negative study results; +: positive (for transgenic mouse assays: at least one examined organ shows an increased mutant/ mutation frequency); ++: majority of results are positive concerning two or more end-points in in vitro studies; (+): study result weakly positive; ±: inconclusive result; nd: no data available; na: not applicable because transgenic assay in this mouse line not done; ––: majority of results are negative concerning two or more end-points; SLRL: sex-linked recessive lethal; somat.: somatic

[a] All data in this table are taken from the Master Table in Appendix 1.

[b] For IARC categories, see footnote b of Table 2.

[c] Gene mutation assays (Ames test; other forward or reverse gene mutation assays in bacteria, e.g. *E. coli* reverse mutation assay; gene mutation assays in mammalian cells, e.g. *Hprt* assay or mouse lymphoma assay; gene mutation assays in fungi, e.g. *Saccharomyces cerevisiae*).

In most cases (16 of 23), the results of the transgenic mutation assays and the mouse spot test were in agreement. This holds for the following compounds: 2-acetylaminofluorene, acrylamide, benzo[*a*]-pyrene, 1,3-butadiene, 1,2-dibromo-3-chloropropane, cyclophosphamide, ethyl methanesulfonate, *N*-ethyl-*N*-nitrosourea, *N*-methyl-*N'*-nitro-*N*-nitrosoguanidine, *N*-methyl-*N*-nitrosourea, 4-nitroquinoline 1-oxide, *N*-nitrosodiethylamine, *N*-nitrosodimethylamine, procarbazine, 4-acetylaminofluorene and *N*-propyl-*N*-nitrosourea. Some compounds (4-acetylaminofluorene, 2-amino-3-methylimidazo[4,5-*f*]quinoline) were positive in the transgenic mutation assays but negative in the spot test, and some compounds (hydrazine sulfate, mitomycin C) were negative in the transgenic mutation assays but positive in the mouse spot test.

The major difference between the transgenic mutation assays and the mouse spot test is that clastogenic substances can also be detected in the mouse spot test. Hydrazine, hydrazine sulfate and trichloroethylene were negative in the transgenic mutation assays. All these substances were, however, positive in the mouse bone marrow micronucleus test, thus revealing the same pattern of results. This is also plausible from the principle of the mouse spot test. In the mouse spot test, there are four possible mechanisms by which the recessive coat colour alleles can be expressed (see section 9.2.1.1), including gene and chromosomal aberrations. Although the chromosomal aberrations also have to survive several mitoses to cause the expression of the recessive allele (Fahrig, 1993), there is evidence that predominantly clastogenic substances might also result in a positive mouse spot test. In contrast, the Big Blue® and Muta™-Mouse transgenic mutation assays detect point mutations and small deletions and insertions (Gossen et al., 1989; Kohler et al., 1991a; Mirsalis et al., 1995).

On the other hand, 4-acetylaminofluorene and 3-amino-3-methylimidazo[4,5-*f*]quinoline were positive in the transgenic mutation assays but negative in the mouse spot test. This may indicate a reduced ability to detect mutations in the mouse spot test, as discussed below. However, only two compounds showed these results.

9.2.1.3 *Predictivity of the transgenic animal mutagenicity assays and the mouse spot test for carcinogenicity*

The sensitivity, specificity and predictivity of carcinogenicity for the transgenic mouse model (Muta™Mouse assay and the Big Blue® mouse assay combined) and the mouse spot test are documented in Table 7. Data on 21 substances (see Table 6) are available on carcinogenicity in mice *and* mutagenic effects in transgenic mice as well as mutagenic effects in the mouse spot test. Two substances (1,2-dichloroethane and trichloroethylene) with inconclusive results in the mouse spot test were not included in the calculation.

Table 7. Characteristics of the Muta™Mouse assay and the Big Blue® mouse assay for predicting mouse carcinogenicity in comparison with the mouse spot test[a]

Term[b]	Calculation for the mouse spot test	Calculation for Muta™Mouse and/or Big Blue® mouse combined[c]
Sensitivity	89% (17/19)	84% (16/19)
Specificity	0 (0/0)	0 (0/0)
Positive predictivity	100% (17/17)	100% (16/16)
Negative predictivity	0 (0/2)	0 (0/3)
Overall accuracy	89% (17/19)	84% (16/19)

[a] Carcinogens with genotoxic and non-genotoxic mechanisms were considered, but not substances without data on carcinogenicity; only data on mice were used. 1,2-Dichloroethane and trichloroethylene were not included in the calculation (inconclusive results in the mouse spot test).

[b] Sensitivity = % of carcinogens with a positive result in the specified test system (STS); specificity = % of non-carcinogens with a negative result in the STS; positive predictivity = % of positive results in the STS that are carcinogens; negative predictivity = % of negative results in the STS that are non-carcinogens; overall accuracy = % of chemicals tested where the STS results agree with the carcinogenicity results.

[c] Judged as positive in transgenic assays if positive in one of the two test systems. For 1,2-dibromo-3-chloropropane and methyl methanesulfonate, the weak positive results in the transgenic assays were judged as positive.

Although the data pool is not sufficient for a comprehensive comparison, there is some indication that no significant differences were detectable between the two test systems.

The results for carcinogens in transgenic mouse assays are compared with those in the mouse spot test and summarized in Table 8. The comparison suggests that results from the mouse spot test and transgenic mouse assays are less complementary than those from the mouse bone marrow assay compared with transgenic mouse assays (see section 9.1). The mouse spot test and the transgenic mouse assays appear to detect the same compounds.

Table 8. Comparison of the results in the mouse spot test and in transgenic mouse assays for carcinogens[a]

	Positive results in the mouse spot test	Negative results in the mouse spot test
Positive results in the Muta™Mouse and/or the Big Blue® mouse assay	15 (80%)	1 (5%)
Negative results in the Muta™Mouse and/or the Big Blue® mouse assay	2 (10%)	1 (5%)

[a] Data taken from Table 6; all substances in this table gave positive results in carcinogenicity studies on mice; weak positive results in transgenic mouse assays were judged as positive; two substances were not included in this table because of inconclusive results in the mouse spot test.

9.2.1.4 *Advantages and disadvantages of both test systems*

a) Sensitivity of the test system

As discussed above (section 9.1.4), the spontaneous mutant frequency in transgenic animals is relatively high. In the mouse spot test, the incidence of spontaneous recessive spots varied between 0.06% and 0.59% (Russell et al., 1981). However, comparing the number of cells and genes at risk at the time of exposure, the mouse spot test is numerically inferior to the transgenic mouse mutation assays. In the mouse spot test, the exposed embryo at gestation day 10 contains about 150–200 melanoblasts, and each melanoblast has four coat colour genes under study (Fahrig, 1977; Russell et al., 1981). In the transgenic Big Blue® mouse, for example, 30–40 copies of the target gene (the constructed λLIZα shuttle vector) are integrated on chromosome 4 of *each* cell of the animal (Kohler et al., 1991a, 1991b). The efficiency of recovery of transgenes (~1%) reduces the number of targets available for analysis.

b) Consideration of animal welfare and cost-effectiveness

To achieve an acceptable sensitivity, a large number of animals are necessary in the mouse spot test. Many pregnant dams have to be in one treatment group to get a sufficient number of surviving F1 animals, since the test substance may induce maternal and/or developmental toxicity. Fahrig (1977) suggested that 30–40 pregnant mice are needed per treatment group for evaluation of spots in the progeny. At least 150 F1 mice are recommended for the concurrent vehicle control (Russell et al., 1981), and at least two dose groups are used (OECD, 1986a). Therefore, the mouse spot test is an expensive type of in vivo test.

In contrast, in transgenic mutation assays, about 20 animals (three dose groups and one concurrent vehicle control group in laboratories that have already established this test system) are recommended per species and gender (Mirsalis et al., 1995; Heddle et al., 2000; Thybaud et al., 2003).

A comparison of both test systems is presented in Table 9.

9.2.1.5 *Conclusions*

Although the mouse spot test is a standard genotoxicity test system according to the OECD guidelines, this system has seldom been used for detection of somatic mutations in vivo in the last decades. This is partly due to considerations of cost-effectiveness and number of animals needed for testing, but also for toxicological considerations. The usefulness of the mouse spot test in toxicology is limited by restrictions in toxicokinetics, sensitivity, target cell/organ and molecular genetics. From the limited data available, it seems that the transgenic mouse assay has several advantages over the mouse spot test and may be a suitable test system to replace the mouse spot test for detection of gene but not chromosome mutations in vivo.

Table 9. Comparison of mouse spot test with the transgenic Big Blue® and Muta™Mouse assays

	Mouse spot test[a]	Transgenic mouse mutation assay[b]
Age restriction	Exposure restricted to embryos on gestation days 9–11	Usually less than 3 months
Toxicokinetics and metabolism	Restrictions in toxicokinetics: test substance reaches the fetal melanoblasts after administration to the dams and absorption of the test substance itself or the toxic metabolites via the placenta	No further barrier like the placenta after absorption and distribution
Target tissue	Restricted to melanoblasts	No tissue restriction; analysis of mutagenic potency in different organs
Type of mutation	Detects 1) gene mutation, 2) large or small deletions, 3) loss of the chromosome carrying the wild-type allele and 4) somatic recombination (marker gene then homozygous)	Detects 1) gene mutation, 2) small deletions or insertions
Dependency of effects on application route	Only systemic effects can be detected; no application route–specific effects	For different routes, systemic as well as local mutagenic effects can be detected
Target gene/cell	Four genes per cell in about 200 melanocytes	About 40 (Big Blue®) or 80 (Muta™Mouse) copies of the transgene per nucleus of each cell of the organism
Number of animals	About 150 pregnant dams per gender per dose (exact number not specified in OECD guideline)	Not more than five animals per gender per dose necessary
Specificity of test system	Discrimination between spots of mutagenic and non-mutagenic origin may be problematical	Identifying and isolating mutated genes with a high specificity

Table 9 (Contd)

	Mouse spot test[a]	Transgenic mouse mutation assay[b]
Characterization of mutations by molecular methods	Less suitable for identification of mutations in DNA analysis due to size of the genes	Detection of the "molecular signature" of a particular mutagen by DNA sequence analysis with standardized methods
Possibility of parallel investigation of several genetic end-points	No combination with other genotoxic end-points suggested	The transgenic mouse assay can be combined with other in vivo genotoxic end-points in the same animal (e.g. micronuclei, chromosomal aberration, UDS, sister chromatid exchange) if results in the transgenic assays are not influenced and vice versa
Endogenous versus foreign target gene	The mouse spot test shows an in situ end-point (expression of the target genes)	Target genes are integrated parts of foreign DNA and consequently no "normal" mutational target
Costs	Expensive type of in vivo test	Uses fewer animals, but the animals are expensive

[a] Fahrig (1977); Styles & Penman (1985); Russell et al. (1981).
[b] Nohmi et al. (2000); RIVM (2000); Health Canada (2004).

9.2.2 *Transgenic animal mutagenicity assay versus* Hprt *and other endogenous genes*

Exogenous reporter genes can be measured in every tissue of transgenic animals as long as sufficient amounts of DNA can be collected, but only a few endogenous genes and tissues are suitable for measuring mutations in vivo. Animal models suitable for measurement at endogenous genes comprise *Hprt*, *Aprt*, *Tk* or *Dlb-1* models. These models detect not only point mutations, frameshifts, small insertions and small deletions, but also intragenic large deletions and loss of heterozygosity (LOH; for *Aprt* and *Tk* genes).

As a comparison of transgenic mouse assays with *Hprt* and other endogenous genes has recently been published elsewhere (RIVM, 2000), it was decided only to update and summarize the discussion in this document (see also Table 10) rather than analysing the individual studies in detail.

Table 10. Comparison of mutation induction in endogenous and exogenous reporter genes in Big Blue® (*lacI*) mice/rats or in Muta™Mouse (*lacZ*) mice[a,b]

Chemical	Splenocytes/splenic lymphocytes			Small intestinal epithelium			Additional remarks	Reference
	Hprt	*lacI*	*lacZ*	*Dlb-I*	*lacI*	*lacZ*		
ENU	++, dr	+					*lacI* mice Mutation spectrum similar in endogenous and exogenous loci	Walker et al. (1994)
ENU	+	+					*lacI* mice Treatment: 1 × i.p.	Skopek et al. (1995)
Ethylene oxide	+	–					B6C3F1 *lacI* mice (Big Blue®) Treatment: inhalation, 366 mg/m^3, 4 weeks Manifestation time: 8 weeks	Sisk et al. (1997); Walker et al. (1997)
BaP	+	++					*lacI* mice	Skopek et al. (1996)
Thiotepa	++	+					*lacI* transgenic rats (Big Blue® F344 rats)	Chen et al. (1998); Casciano et al. (1999)
DMBA	++, dr	+					*lacI* transgenic rats (Big Blue® F344 rats)	Manjanatha et al. (1998); Casciano et al. (1999)
CP	+	–					*lacI* mice Treatment: 1 × i.p.	Walker et al. (1999a)

Table 10 (Contd)

Chemical	Splenocytes/splenic lymphocytes			Small intestinal epithelium			Additional remarks	Reference
	Hprt	*lacI*	*lacZ*	*Dlb-I*	*lacI*	*lacZ*		
(contd)							Manifestation time: 6 weeks	
N-OH-AAF	+	+					Big Blue® F344 rats, i.p., 1, 2, 4 repeats	Chen et al. (2001a)
							Manifestation time: 6 weeks	
MNU	++, dr	+					*lacI* mice (Big Blue®)	Monroe et al. (1998)
BaP	+	+					Treatment: 1 × i.p.	
PhIP				+	+		*lacI* mice (Big Blue®)	Zhang et al. (1996)
							Treatment: oral, 30, 60, 90 days	
ENU, i.p.	nt	nt		+	+		*lacI* mice	Tao et al. (1993a)
X-ray	+	+		+	–		Treatment: 1 × i.p.	
MMS							*lacI* mice	Tao et al. (1993b)
				–	–		Treatment: 1 × i.p.	
				+	+		Treatment: subacute	
MNU				++, dr		++, dr	F1 (Muta™Mouse × SWR) mice	Cosentino & Heddle (1999)
BrdU				+, dr		+	Treatment: 1 × p.o.	
EMS				+, ss		±, ns	Manifestation time: 2 weeks	
MMS				+, dr		+, dr		

Table 10 (Contd)

Chemical	Splenocytes/splenic lymphocytes			Small intestinal epithelium			Additional remarks	Reference
	Hprt	*lacI*	*lacZ*	*Dlb-l*	*lacI*	*lacZ*		
BaP				+, dr		+, dr	*(see above)*	Cosentino & Heddle (1999)
MMC				+, ss		±, ns		
ENU	+, dr		+, dr	+, dr		+, dr	*lacZ*$^{+/0}$ / *Dlb-1*$^{a/b}$ mice	van Delft et al. (1998)
MNU	+		+	+		+	Treatment: 1 × i.p. or i.p. split dose 5 × 1/5 (only ENU)	
EMS	–		–	–		–	Manifestation time: 7 weeks	
ENU	+		++, dr	+, dr		++, dr	F1 (Muta™Mouse × SWR) mice; chronic exposure via drinking-water (94 µg ENU/ml) or diet (40 mg BaP/kg); *lacZ* linear accumulation of mutations, endogenous locus non-linear	Cosentino & Heddle (2000)
BaP				+		+, dr		

BaP = benzo[*a*]pyrene; BrdU = 5-bromo-2'-deoxyuridine; CP = cyclophosphamide; DMBA = 7,12-dimethylbenz[*a*]anthracene; EMS = ethyl methanesulfonate; ENU = *N*-ethyl-*N*-nitrosourea; MMC = mitomycin C; MMS = methyl methanesulfonate; MNU = *N*-methyl-*N*-nitrosourea; N-OH-AAF = *N*-hydroxy-2-acetylaminofluorene; PhIP = 2-amino-1-methyl-6-phenylimidazo[4,5-*b*]pyridine; i.p. = intraperitoneal; p.o. = per os (by mouth); nt = not tested; dr = dose-related, ss = statistically significant, ns = not statistically significant

[a] Adapted from RIVM (2000).

[b] Studies were not evaluated. The test outcomes in the table represent the conclusion as given in the papers, i.e. –: no treatment-related increase; ±: outcome inconclusive; +: treatment-related increase; ++: increase in mutant/mutation frequency more pronounced relative to others in the same series of experiments.

In the development and use of transgenic assays, it is assumed that mutations in the exogenous reporter genes accurately reflect mutations at endogenous loci. However, the sequences and location in the genome differ between exogenous and endogenous loci. The types and frequencies of mutations detected at a locus, whether transgenic or endogenous, depend on many factors, including the location of the gene, the selective system used and the sequence of the DNA within the gene. Many differences are known between endogenous loci with respect to spontaneous mutation rate, induced mutant frequency and mutation spectrum. The *Hprt* locus, for example, has a much lower spontaneous mutant frequency than *Dlb-1*, but a much higher mutant frequency than *Oua* (ouabain resistance). The latter is a dominant mutation in an essential gene, in which only base substitutions are detectable, and these at only a few base pairs. In contrast, *Hprt* is a non-essential gene, present in only one functional copy per cell, so a wide variety of base substitutions and deletions even of the whole gene and beyond are detectable. Even larger deletions and rearrangements corresponding to LOH are detectable at *Tk* and *Aprt*. In contrast, most transgenic systems detect base substitutions or deletions within the gene or vector array. Only the Spi$^-$ assay (deletions up to 10 000 bp) and the *lacZ* plasmid assay can detect larger deletions. The most common site of base substitution in mammalian cells, CpG, is relatively rare in endogenous genes, but is quite frequent in the transgenes. The nature of the selection and the structure of the protein influence the number of mutable sites within the gene, which also influences the mutation rate. Finally, mutants at some loci, including *Hprt*, are at a selective disadvantage, which reduces the mutant frequency. Thus, it is not surprising that differences are observed among loci. Indeed, it is surprising that the *cII* and *lacZ* transgenes, which differ in size by 10-fold, have such similar mutant frequencies. Mutation spectra for base substitutions are quite similar for *Hprt* and the transgenes. Further, prokaryotic DNA is heavily methylated, non-transcribed and embedded in bacteriophage DNA. Transgenes are usually present in multiple tandem copies. Comparisons of mutations in endogenous genes and transgenes in the same tissue are valuable to evaluate the use of transgenic animals in toxicity testing.

9.2.2.1 Description of endogenous gene animal models

a) The *Hprt* rodent model

Hprt (hypoxanthine-guanine phosphoribosyltransferase) is an endogenous gene present in all tissues, but mutant selection is predominantly performed in splenocytes or human peripheral T lymphocytes or any tissue from which viable cells can be subcloned. *Hprt* is a non-essential enzyme for cells in culture. Mutants are selected by culture in the presence of 6-TG, which is a substrate for the enzyme. It is converted into the corresponding monophosphate, which is in turn toxic to cells. *Hprt* mutants have lost this enzyme activity and can grow in medium containing 6-TG. The *Hprt* gene is located on the X chromosome and spans 32 kb in rodent cells and 46 kb in human cells. It has a coding region of 657 bp (Skopek et al., 1995). The *Hprt* data are complicated by the fact that the mutant frequency varies with time after treatment, and this time response is age-dependent (Walker et al., 1999b).

The *Hprt* model detects small mutations, intragenic deletions much larger than those detected by *lacI* and *lacZ* and deletions extending beyond the gene. However, very large deletions and LOH are not revealed, as essential genes may be deleted from the single, functional X chromosome. Consequently, if large deletions enclose adjacent genes that are essential for cell survival, these cells will not survive.

b) The *Aprt* mouse model

The *Aprt* (adenine phosphoribosyl transferase) gene codes for a protein that converts adenine into adenosine monophosphate. The human *APRT* gene is located on chromosome 16 and is 2.6 kb in length; the mouse *Aprt* gene is located near the telomere on chromosome 8. In the C57BL/6 *Aprt* mouse model, the gene was knocked out by homologous recombination in embryonic stem cells; a part of the promoter region as well as the ATG start codon were deleted (Engle et al., 1996; van Sloun et al., 1998). Because of the recessive nature of *Aprt* mutations, heterozygous *Aprt* mice are used for genotoxicity testing. The *Aprt* model detects small mutations, intragenic large deletions and LOH. This assay can be used in any tissue from which viable cells can be cloned.

c) The *Tk* mouse model

The *Tk* (thymidine kinase) gene is an autosomal gene, telomeric on chromosome 11, which participates in pyrimidine salvage by converting thymidine to thymidine monophosphate. Because heterozygous cells are also sensitive to selective agents, one *Tk* allele was inactivated by homologous recombination in embryonic stem cells of the 129 mouse and subsequently backcrossed to C57BL/6. A novel gene mutation assay using this gene was developed (Dobrovolsky et al., 1999, 2005) and is commercially available. Because of the recessive nature of *Tk* mutations, heterozygous *Tk* mice are used for genotoxicity testing. The advantage of the *Tk* model is its sensitivity for large deletions, large chromosomal alterations and LOH. The disadvantage is that the *Tk* model uses 5-bromo-2'-deoxyuridine as a selective agent; 5-bromo-2'-deoxyuridine is itself a mutagen and may contribute to the background mutation frequency (RIVM, 2000). It is thought that the high, chronic dose of 5-bromo-2'-deoxyuridine used for selection prevents survival of cells that could be mutated by the selective agent. This assay can be used in any tissue from which viable cells can be cloned.

d) *Dlb-1* assay

The *Dlb-1* assay allows scoring of mutations in the small intestine (and possibly in the colon) of the mouse (Winton et al., 1988). *Dlb-1* is a polymorphic gene on chromosome 11 with two alleles. *Dlb-1*^{b}, present in most mouse strains, leads to expression of a binding site for the lectin *Dolichos biflorus* agglutinin in intestinal epithelium, whereas *Dlb-1*^{a}, present in SWR mice and very few other strains, determines the expression in vascular epithelium. The assay is based on recognition of mutations affecting the *Dlb-1*^{b} gene of heterozygotic *Dlb-1*^{a}/*Dlb-1*^{b} mice. The *Dlb-1*^{a}/*Dlb-1*^{b} epithelial cells stain dark brown, and mutant cells (which have no lectin binding sites) appear as unstained vertical stripes on the villi (Winton et al., 1990; Tao et al., 1993a, 1993b). Since the *Dlb-1* mutations have not yet been sequenced, the molecular nature of the mutations has not been determined in DNA sequences.

9.2.2.2 Comparative studies

a) Studies comparing mutational response of transgenic animals with the *Hprt* gene

Comparison of the mutational response of the *lacI* transgene in Big Blue® mouse with that of the native *Hprt* gene in the same treated animals has been performed for a number of substances: benzo[*a*]pyrene, cyclophosphamide, 7,12-dimethylbenz[*a*]anthracene, *N*-ethyl-*N*-nitrosourea, ethylene oxide, *N*-hydroxy-2-acetylaminofluorene, *N*-methyl-*N*-nitrosourea, thiotepa and X-ray (see Table 10).

In the first of these studies, the frequency and spectrum of mutations induced at the *Hprt* and *lacI* loci of splenic lymphocytes were defined and compared following acute exposures of young male *lacI* transgenic mice to an experimental direct-acting alkylating agent, *N*-ethyl-*N*-nitrosourea. The resulting data indicated that the average induced mutant frequencies (i.e. induced mutant frequency = treatment mutant frequency minus background mutant frequency) and the types of mutations produced by *N*-ethyl-*N*-nitrosourea in *lacI* and *Hprt* were similar; however, the *lacI* mutation assay was less sensitive than the *Hprt* assay for detecting increases in mutant frequency following *N*-ethyl-*N*-nitrosourea treatment (Walker et al., 1999a). In contrast, Skopek et al. (1996) found that the *Hprt* assay was less sensitive than the *lacI* assay for the detection of benzo[*a*]pyrene-induced mutations. In a further study using cyclophosphamide, under the treatment conditions used, cyclophosphamide-induced mutations in splenic lymphocytes were detectable in the *Hprt* gene but not the *lacI* transgene of this non-target tissue for cyclophosphamide-induced cancer (Walker et al., 1999a). However, using multiple dosing protocols that are more consistent with those currently recommended (Thybaud et al., 2003), cyclophosphamide has been shown to yield positive results in bone marrow (Myhr, 1991; Hoorn et al., 1993).

In Big Blue® mice exposed to ethylene oxide at 366 mg/m^3, the *lacI* mutant frequency in the lung (carcinogenicity target organ) was significantly increased at 8 weeks post-exposure but not in spleen and bone marrow (Sisk et al., 1997; Walker et al., 1997). The occurrence of a detectable mutational response at *Hprt* but not at the *lacI* transgene in spleen cells is likely due to the mechanism of action of ethylene oxide–induced mutation. Molecular characterization of ethylene oxide–induced *HPRT* mutations in diploid human fibroblasts in vitro (Bastlová et al., 1993) has indicated that as many as 50% of the *HPRT* mutations induced by ethylene oxide are large deletions, often involving the loss of the entire *HPRT* gene. If this is the case in vivo as well as in vitro, then differences in the *Hprt* and *lacI* mutant frequencies may be due to the recovery of large deletions as part of the *Hprt* mutant frequency, but lack of these events at the *lacI* transgene (Gossen et al., 1995).

The mutant frequencies of the *lacI* transgene of Big Blue® rats were compared with those of the endogenous *Hprt* using 7,12-dimethylbenz[*a*]anthracene administration by gavage. The *Hprt* and *lacI* genes differed with respect to the kinetics of mutant induction, the magnitudes of both the spontaneous and 7,12-dimethylbenz[*a*]anthracene-induced mutant frequency response and the ability to detect mutants induced by 7,12-dimethylbenz[*a*]anthracene exposure. High spontaneous mutant frequency and variability associated with detecting mutations in the *lacI* gene contributed to the reduced sensitivity in the assay. In particular, mutant frequencies in the animals treated with the low dose of 7,12-dimethylbenz[*a*]-anthracene were significantly higher in the *Hprt* gene than those in the control animals, while no such induction was found in the *lacI* gene in nearly all experiments (Manjanatha et al., 1998). In spleen, failure to detect an increase in mutant frequency in the *lacI* gene could also result from the higher background mutant frequency in the *lacI* gene than in the *Hprt* gene.

The induced mutant frequency of thiotepa-treated Big Blue® rats was 2.8-fold greater in the *lacI* gene than in the *Hprt* gene, although the *Hprt* gene recovered large deletions not found among the *lacI* gene. The authors discussed two reasons for these differences: transcription-coupled DNA repair in the *Hprt* gene and the targeting of base pair substitutions to G:C base pairs in the *lacI* transgene. However, comparing the fold increase in mutant frequency from treated animals relative to the controls, the increase was more pronounced in the *Hprt* gene (12-fold versus 4-fold) (Chen et al., 1998; Casciano et al., 1999).

N-Hydroxy-2-acetylaminofluorene administered to Big Blue® rats in multiple doses caused increased mutant frequencies in both the *Hprt* and *lacI* genes of spleen lymphocytes and about 10-fold more *lacI* mutations in the liver than in spleen lymphocytes (Chen et al., 2001a; see also section 11.3.3). Sequence analysis showed significant differences in the patterns of base pair substitution and frameshift mutation between liver and spleen *lacI* mutants and between spleen lymphocyte *lacI* and *Hprt* mutants. Twelve per cent of mutants from treated rats had major deletions in the *Hprt* gene, whereas no corresponding incidence of large deletions was evident among *lacI* mutations (see also types of mutation in section 6.1.1). The differences between *N*-hydroxy-2-acetylaminofluorene mutation in the endogenous gene and transgene can be partially explained by the structures of the two genes. For example, among the *lacI* frameshifts are four deletions of CG/GC in the DNA sequences GCGC. This frameshift mutation did not occur in the *Hprt* gene, but only one GCGC sequence is found in the *Hprt* coding region compared with 22 in the *lacI* coding region (Chen et al., 2001b).

The results of these studies indicate that the frequencies of *Hprt* and *lacI* mutants induced by various mutagenic carcinogens are rarely the same, but depend on the nature of the target gene. For instance, agents that are mainly point mutagens generally produce higher mutant frequencies in the *lacI* gene than in the *Hprt* gene, because the *lacI* gene has a larger target for point mutation, especially for point mutation at G:C base pairs (261 recoverable base pair mutations in the *lacI* gene and 149 in the *Hprt* gene) (Chen et al., 1998, 2001a).

b) Studies comparing mutational response of transgenic animals with the *Dlb-1* locus

Another set of studies has focused on the *Dlb-1* locus. Exposure to *N*-ethyl-*N*-nitrosourea was used to compare the frequencies of *lacI* mutants in half of the small intestine with the frequencies of the host *Dlb-1^b to Dlb-1^a* mutations induced in the other half. The *lacI* transgene and the endogenous *Dlb-1* locus responded similarly after intraperitoneal treatment with *N*-ethyl-*N*-nitrosourea but responded differently after treatment with X-rays (Tao et al., 1993a). This difference is probably due to the fact that X-rays produce predominantly double-stranded DNA breaks and, through these, deletions that are not detected in *lacI* transgenic mice. In a further study, methyl methanesulfonate produced no significant increase in mutations at either locus. Subacute treatments produced low but significant increases in mutant frequency at both loci (Tao et al., 1993b).

A further study compared the effects of diverse mutagens at the *lacZ* transgene and *Dlb-1* locus in vivo (Cosentino & Heddle, 1999). Benzo[*a*]pyrene, 5-bromo-2'-deoxyuridine, methyl methanesulfonate, ethyl methanesulfonate, *N*-ethyl-*N*-nitrosourea, mitomycin C and *N*-methyl-*N*-nitrosourea were all given by gavage to F1/ Muta™Mouse × SWR mice, and the mutations were quantified 2 weeks after the end of treatment. Although each mutagen produces a distinct spectrum of mutations, resulting from the specificity of DNA binding and type of DNA repair involved, all of the agents induced similar mutant frequencies at the *Dlb-1* locus and at the *lacZ* transgene, although a higher background frequency was observed at the *lacZ* transgene.

During chronic mutagen exposure, mutations at the transgene accumulate linearly with time (i.e. in direct proportion to the dose received). In contrast, mutations at the endogenous gene are much less frequent than those of the transgene early in the exposure period, and the accumulation is not linear with time (Shaver-Walker et al., 1995), but rather accelerates as the exposure continues. This mutational response is not limited to one genetic background or to one locus, one tissue or one mutagen, but is a more general event (Cosentino & Heddle, 2000). This could reflect a difference in repair efficiency at low damage levels.

c) Studies comparing mutational response of transgenic animals with the *Dlb-1* locus and *Hprt* locus together

Van Delft et al. (1998) studied alkylation-induced mutagenesis 1) in *lacZ* and *Hprt* in spleen cells and 2) in *lacZ* and *Dlb-1* in small intestine from F1/Muta™Mouse × SWR mice 7 weeks after single intraperitoneal injection of *N*-ethyl-*N*-nitrosourea, *N*-methyl-*N*-nitrosourea and ethyl methanesulfonate (see Table 10). With *N*-ethyl-*N*-nitrosourea, split-dose treatment was also performed (1 × 50 mg/kg of body weight or 5 × 10 mg/kg of body weight with a 1- or 7-day interval). Except for ethyl methanesulfonate, a dose-related mutagenic effect was seen in *lacZ* and *Dlb-1*. Furthermore, results suggest that mutagenic effects of fractionated doses are generally additive. In most cases, the induction factor (ratio treated over controls) for mutations in *lacZ* was lower than that for *Hprt* and *Dlb-*

1, presumably due to a higher background in *lacZ* and/or a lower mutability of *lacZ*. The authors concluded that the general concordance between data for *lacZ* and the endogenous genes indicates that *lacZ* transgenic mice are a suitable model to study induction of gene mutations in vivo.

In a further study comparing the *Dlb-1* locus and the *lacZ* transgene from the Muta™Mouse in the small intestine and the *Hprt* locus and the *lacZ* transgene in splenocytes, comparisons were made in both tissues after acute and chronic exposure to *N*-ethyl-*N*-nitrosourea and in the small intestine to benzo[*a*]pyrene. All comparisons showed that during chronic exposures, mutations at the transgene accumulate linearly with increasing duration of exposure, whereas induced mutations of the endogenous gene initially accumulate at a slower rate (Cosentino & Heddle, 2000). Identical results were reported by Shaver-Walker et al. (1995). This phenomenon could reflect a difference in repair efficiency at low damage levels.

In a comparative study of *Hprt*, *lacI* and *cII/cI* as mutational targets for *N*-methyl-*N*-nitrosourea and benzo[*a*]pyrene in Big Blue® mice, the order of mutation assay sensitivity was *Hprt* > *lacI* > *cII/cI* with *N*-methyl-*N*-nitrosourea and *Hprt* ≈ *lacI* > *cII/cI* for benzo[*a*]pyrene (Monroe et al., 1998).

9.2.2.3 Conclusion

Despite differences in the mutational properties of the various model mutagens, the response of the exogenous loci (*lacI*, *lacZ* transgene) and the endogenous loci (*Dlb-1*, *Hprt*) were generally qualitatively similar following acute treatments. Several studies suggest that the lower spontaneous mutant frequency in the endogenous genes may provide enhanced sensitivity under such conditions. However, comparisons of transgenes and endogenous genes are difficult because of differences between the optimal experimental protocols for the different types of genes; in the neutral transgenes, sensitivity for the detection of mutations is increased, with administration times that are longer than those currently recommended.

9.3 Transgenic animal mutagenicity assays and indirect measure of DNA damage using UDS in vivo assay

The transgenic animal mutagenicity assays were also compared with the in vivo rat liver UDS assay. Dean et al. (1999) reviewed the data for 12 rodent carcinogens, all of which were detected by either Big Blue® or Muta™Mouse. Of these, seven were tested using the UDS assay and found to be negative, whereas three were negative in the in vivo micronucleus test. Dean et al. (1999) observed that for substances applied to the skin, orally dosed or inhaled and which may not reach either the bone marrow or liver in active form, then conducting a transgenic animal mutagenicity assay using an appropriate tissue may be a more suitable approach. Although a comparison was not done in the present document, these results suggest that transgenic animal mutagenicity assays exhibit superior predictivity compared with the UDS test. This is not unexpected, since the UDS assay measures initial DNA damage and is recognized as an indicator test for genotoxicity.

9.4 Results of transgenic animal mutagenicity assays compared with results of genotoxicity assays in vitro

Using data from the Master Table (Appendix 1), a comparison was made between results found in Muta™Mouse and Big Blue® assays (combined) and three different end-points in genotoxicity in vitro:

1) gene mutation (Ames test; other forward or reverse gene mutation assays in bacteria, e.g. *E. coli* reverse mutation assay; gene mutation assays in mammalian cells, e.g. *Hprt* assay or mouse lymphoma assay; gene mutation assays in fungi, e.g. in *Saccharomyces cerevisiae*);
2) chromosomal aberration (cytogenetic assay in mammalian cells; micronucleus assay in mammalian cells; cytogenetic assay and/or testing of aneuploidy in fungi, e.g. *S. cerevisiae*); and
3) direct or indirect measures of DNA damage (DNA damage in bacteria measured, for example, by the rec-assay or the SOS-umu-test; mitotic recombination assay in *S. cerevisiae*; assays on DNA adducts in mammalian cells; sister chromatid exchange assay in mammalian cells; DNA damage and repair, UDS in mammalian cells)

to see if studies on genotoxicity in vitro, especially gene mutation, are in agreement with results in the transgenic animal mutagenicity assay, which detects gene mutations including small deletions and insertions. In Table 11, results on transgenic animal mutagenicity assays and these three different end-points in genotoxicity in vitro are tabulated for each substance. A summary of this comparison is given in Table 12.

9.4.1 Gene mutation

Comparing in vitro gene mutation with transgenic animal assays points to an agreement between both test systems (see Table 12). Nearly all substances (36 of 42) positive for any of three transgenic animal mutagenicity assays (marked TG+ in Table 12 if TG(+) up to TG+++ in Table 11) also gave positive results in studies on gene mutation in vitro; three substances (benzene, phenobarbital, urethane) showed inconclusive results in vitro. This tendency is independent of the number of test systems available for testing mutagenicity in transgenic animals (TG+++ for three test systems in Table 11 and TG+ for one). Only three substances (e.g. asbestos crocidolite) gave negative results in vitro but a positive result in the transgenic animal mutagenicity assay. Asbestos is carcinogenic in the lung after inhalation, and the in vitro assays on gene mutation were apparently unsuitable for detection of genotoxic mechanisms (IARC, 1987b, 1987c), in contrast to the transgenic assay on Big Blue® mice, also using the inhalation exposure route (see also section 10.2).

Five of 13 substances with negative results in transgenic animal mutagenicity assays were also negative in vitro, and 2 of 13 showed inconclusive in vitro results. However, the remaining six substances (bromomethane, 2,6-diaminotoluene, 1,2-dichloroethane, hydrazine, mitomycin C, trichloroethylene) had positive results in vitro. These differences may be the result of suboptimal experimental design (i.e. not meeting the standards currently recommended for transgenic studies) (Thybaud et al., 2003).

9.4.2 Chromosomal aberration

The database for this comparison is limited compared with the other end-points (no data in vitro on 9 of 55 substances; see Table

Table 11. Comparison of results of transgenic animal mutagenicity assays (TG) and genotoxicity assays *in vitro* (data taken from the Master Table, Appendix 1)

Substance	Results of TG	Genotoxicity *in vitro*					
		Gene mutation		Chromosomal aberration		Direct or indirect measure for DNA damage	
		Results	In agreement with TG	Results	In agreement with TG	Results	In agreement with TG
2-Acetylaminofluorene	++	++	Yes	+	Yes	?	Inconclusive
4-Acetylaminofluorene	+	++	Yes	nd	na	--	No
Acrylamide	+	++	Yes	++	Yes	++	Yes
Aflatoxin B1	++	++	Yes	+	Yes	++	Yes
Agaritine	(+)	(+)	Yes	nd	na	nd	na
4-Aminobiphenyl	+	++	Yes	nd	na	++	Yes
2-Amino-3,4-dimethyl-imidazo[4,5-*f*]quinoline	+	++	Yes	++	Yes	++	Yes
2-Amino-3,8-dimethyl-imidazo[4,5-*f*]-quinoxaline	++	++	Yes	?	Inconclusive	++	Yes
2-Amino-1-methyl-6-phenylimidazo-[4,5-*b*]-pyridine	++	++	Yes	+	Yes	++	Yes

Table 11 (Contd)

Substance	Results of TG	Genotoxicity *in vitro*					
		Gene mutation		Chromosomal aberration		Direct or indirect measure for DNA damage	
		Results	In agreement with TG	Results	In agreement with TG	Results	In agreement with TG
2-Amino-3-methyl-imidazo[4,5-*f*]quinoline	++	++	Yes	?	Inconclusive	++	Yes
ortho-Anisidine	+	++	Yes	+	Yes	++	Yes
Asbestos crocidolite	+	––	No	++	Yes	––	No
Benzene	+	?	Inconclusive	–	No	––	No
Benzo[*a*]pyrene	++	++	Yes	++	Yes	++	Yes
Bromomethane	–	++	No	+	No	++	No
1,3-Butadiene	++	–	No	nd	na	?	Inconclusive
Chorambucil	+	++	Yes	+	Yes	++	Yes
Chloroform	–	?	Inconclusive	–	Yes	?	Inconclusive
Cyclophosphamide	++	++	Yes	++	Yes	++	Yes
2,4-Diaminotoluene	+	++	Yes	+	Yes	++	Yes
2,6-Diaminotoluene	–	++	No	++	No	–	Yes
1,2-Dibromoethane	+	++	Yes	++	Yes	++	Yes

Table 11 (Contd)

Substance	Results of TG	Genotoxicity *in vitro*					
		Gene mutation		Chromosomal aberration		Direct or indirect measure for DNA damage	
		Results	In agreement with TG	Results	In agreement with TG	Results	In agreement with TG
1,2-Dibromo-3-chloro-propane	(+)	++	Yes	++	Yes	++	Yes
1,2-Dichloroethane	–	++	No	++	No	++	No
Di-(2-ethylhexyl) phthalate	–	––	Yes	––	Yes	––	Yes
5-(*p*-Dimethylamino-phenylazo)benzo-thiazole	+	+	Yes	nd	na	nd	na
6-(*p*-Dimethylamino-phenylazo)benzo-thiazole	+	+	Yes	nd	na	nd	na
7,12-Dimethylbenz[*a*]-anthracene	+++	++	Yes	+	Yes	++	Yes
Ethylene oxide	+	++	Yes	++	Yes	++	Yes
Ethyl methanesulfonate	+	++	Yes	++	Yes	++	Yes
N-Ethyl-*N*-nitrosourea	++	++	Yes	++	Yes	++	Yes
Heptachlor	–	––	Yes	nd	na	––	Yes

Table 11 (Contd)

Substance	Results of TG	Genotoxicity *in vitro*					
		Gene mutation		Chromosomal aberration		Direct or indirect measure for DNA damage	
		Results	In agreement with TG	Results	In agreement with TG	Results	In agreement with TG
Hydrazine	–	++	No	?	Inconclusive	++	No
(+)-Limonene	–	–	Yes	–	Yes	nd	na
Methyl methanesulfonate	(+)	++	Yes	++	Yes	++	Yes
N-Methyl-*N'*-nitro-*N*-nitrosoguanidine	+	++	Yes	++	Yes	++	Yes
4-(Methylnitrosamino)-1-(3-pyridyl)-1-butanone	+	++	Yes	nd	na	+	Yes
N-Methyl-*N*-nitrosourea	++	++	Yes	++	Yes	++	Yes
Mitomycin C	–	++	No	++	No	++	No
4-Nitroquinoline 1-oxide	+	++	Yes	++	Yes	++	Yes
N-Nitrosodiethylamine	+	++	Yes	+	Yes	++	Yes
N-Nitrosodimethylamine	+++	++	Yes	+	Yes	++	Yes
N-Nitrosodi-*n*-propyl-amine	+	++	Yes	+	Yes	+	Yes
Phenobarbital	(+)	?	Inconclusive	?	Inconclusive	––	No
Procarbazine	+	++	Yes	–	No	+	Yes

Table 11 (Contd)

Substance	Results of TG	Genotoxicity *in vitro*					
		Gene mutation		Chromosomal aberration		Direct or indirect measure for DNA damage	
		Results	In agreement with TG	Results	In agreement with TG	Results	In agreement with TG
β-Propiolactone	+	++	Yes	+	Yes	++	Yes
N-Propyl-*N*-nitrosourea	+	++	Yes	nd	na	?	Inconclusive
Quinoline	+	+	Yes	+	Yes	++	Yes
Sodium saccharin	–	?	Inconclusive	?	Inconclusive	?	Inconclusive
Tamoxifen	+	––	No	+	Yes	++	Yes
2,3,7,8-Tetrachloro-dibenzo-*p*-dioxin	–	––	Yes	+	No	?	Inconclusive
Tetrachloromethane	–	–	Yes	?	Inconclusive	+	No
Trichloroethylene	–	(+)	No	?	Inconclusive	–	Yes
Tris(2,3-dibromopropyl)-phosphate	+	++	Yes	?	Inconclusive	++	Yes
Urethane	++	?	Inconclusive	+	Yes	++	Yes

Table 11 (Contd)

Legend for transgenic animal assays (column 2):

+: positive in the Muta™Mouse or the Big Blue® mouse or the Big Blue® rat; ++: positive in two out of these three test systems; +++: positive in all three test systems; −: negative in one of the three test systems (no substance tested in two test systems with exclusively negative results); (+): weak positive result in the Big Blue® mouse assay or the Muta™Mouse assay

Legend for data on genotoxicity in vitro (columns 3–8)

+: positive concerning one genotoxic end-point; ++: majority of results are positive concerning two or more end-points; (+): study result weakly positive or weak positive effects in two or more tested end-points; ?: equivocal results concerning two or more end-points, inconclusive result concerning one end-point; −: negative concerning one tested end-point; −−: majority of results are negative concerning two or more end-points; nd: no data available; na: not applicable

Table 12. Summary of comparison of results of transgenic animal mutagenicity assays (TG) and in vitro genotoxicity assays

TG assays		Results of in vitro genotoxicity assay											
Results	Number of substances	Gene mutation				Chromosomal aberration				Direct or indirect measure for DNA damage			
		+	−	Inconc.	nd/na	+	−	Inconc.	nd/na	+	−	Inconc.	nd/na
TG+	42	36	3	3	0	28	2	4	8	32	4	3	3
TG−	13	6	5	2	0	5	3	4	1	5	4	3	1

TG+: positive in at least one transgenic animal assay in Table 11 (weak positive results were judged as positive), TG−: negative results in transgenic animal assays

Legend for data on genotoxicity in vitro:

+: positive (marked by +, ++ or (+) in Table 11); −: negative (marked by − or −− in Table 11); Inconc.: inconclusive results (marked by ? in Table11); nd/na: no data available, comparison not applicable

12). Although the transgenic mutation assays are less suitable for detection of clastogenic effects, an agreement was seen in most cases between results in transgenic animals and chromosomal aberration assays in vitro (see also comparison of transgenic animal mutagenicity assays with the micronucleus test in section 9.1). This might be due to the fact that compounds that exclusively induce point mutations or chromosomal aberrations are not available, although a preference for one of these end-points may exist. Apparently, most mutagens induce point mutations as well as chromosome breakage. Five substances with negative results in transgenic animal mutagenicity assays gave positive results in in vitro assays on chromosome mutation: bromomethane, 2,6-diaminotoluene, 1,2-dichloroethane, mitomycin C and 2,3,7,8-tetrachlorodibenzo-*p*-dioxin (Table 11). This discrepancy might be due to the predominantly clastogenic activity of the test substance (e.g. mitomycin C) or simply a suboptimal experimental design of the transgenic animal mutagenicity study.

9.4.3 Direct or indirect measure of DNA damage

The correspondence between the results of transgenic animal mutagenicity studies and the results of in vitro studies other than gene and chromosomal aberration assays is very clear (Table 12). Contradictory results are documented for only five substances with negative outcome in the transgenic animal mutagenicity assay but positive results in vitro (bromomethane 1,2-dichloroethane, hydrazine, mitomycin C and tetrachloromethane) and four substances with positive results in the transgenic animal mutagenicity assay but negative outcome in vitro (4-acetylaminofluorene, asbestos crocidolite, benzene and phenobarbital) (see Table 11).

9.4.4 Conclusion

The results of genotoxicity assays in vitro are in good agreement with the results of transgenic animal mutation assays.

10. TRANSGENIC ASSAYS AND CARCINOGENICITY TESTING

10.1 Comparison of target organs in carcinogenicity studies with target organs in transgenic animal mutation assays

A major advantage of the transgenic mouse/rat mutation assay compared with other in vivo mutagenicity tests is that mutagenic events in any organ can be detected. Therefore, studies have been undertaken to investigate whether the transgenic animal mutagenicity assay can be used to predict target organs in carcinogenicity studies.

10.1.1 Pattern of target organs

In a collaborative study, target organs in transgenic animals (Muta™Mouse) were compared with target organs in carcinogenicity studies for several substances (Suzuki et al., 1999a). *N*-Nitrosodi-*n*-propylamine, propylnitrosourea, 7,12-dimethylbenz[*a*]-anthracene and 4-nitroquinoline 1-oxide were administered by intraperitoneal injection, whereas procarbazine was administered orally. The mutant frequency in different organs was determined after 7, 14 and 28 days. Organs analysed included known target organs for carcinogenicity as well as non-target organs. All chemicals studied were found to cause an increase in *lacZ* mutant frequency in their carcinogenesis target organs. Some non-target organs for cancer, however, also showed positive responses, although the mutant frequencies were generally lower than those in the target organs.

In the study of Suzuki et al. (1999a), organs with a high proliferation rate, such as bone marrow, stomach (mucosa) and colon (mucosa), tended to show a higher mutant frequency than other organs. This is consistent with the view that mutations are more prone in rapidly dividing than in slowly dividing cells. Similar results were presented by Nagao et al. (1998), comparing results from their carcinogenicity study with 2-amino-3,4-dimethylimidazo-[4,5-*f*]quinoline with those from a mutagenicity study under similar experimental design and using the same strain and the same gender of Big Blue® mice (C57BL/6) (Suzuki et al., 1996b).

In order to extend these comparisons, in this document, additional studies have been analysed in Muta™Mouse and in Big Blue® mice or rats (see Table 13) where:

- the same route of exposure and the same species were used for the carcinogenicity study and transgenic animal mutagenicity assay; and
- three or more organs have been investigated in transgenic animal mutagenicity assays.

In the case of lymphomas, the target organ of mutagenicity was assumed to be one of the lymphatic organs (e.g. spleen, thymus); in the case of leukaemia, the target organ was the bone marrow. The results of the transgenic animal mutagenicity assays were taken from primary references; the information on carcinogenicity was obtained from reviews (see Master Table, Appendix 1).

When analysing Table 13, one has to bear in mind that the animal strains investigated in the carcinogenicity studies were most often different from those in the transgenic animal mutagenicity assays. In addition, dose levels and exposure duration may have differed considerably. Furthermore, only a relatively small number of carcinogenicity studies performed according to current guidelines (e.g. OECD Test Guideline 451: OECD, 1981) were available. In some studies, the number of animals was small, or sometimes not all organs were examined histopathologically. With respect to the transgenic assay protocol, the experimental conditions may not have been optimal, a fact that may lead to false-negative results (e.g. see hydrazine). Transgenic animal mutagenicity assays with negative outcome have been evaluated with respect to their validity (see section 6.2 and Table 18). Furthermore, limitations occurred if target organs in carcinogenicity studies were not examined in transgenic animal mutagenicity assays (in the case of ethyl methanesulfonate). Despite all these limitations, some general trends could be observed:

- The organs with the highest mutant frequencies are not necessarily the target organ for cancer.

Table 13. Target organs in carcinogenicity studies compared with those in transgenic assays[a]

Substance	Species	Route	Target organs in carcinogenicity studies	Organs examined in transgenic animal assays					
				Muta™-Mouse	Muta-genicity	Agree-ment[b]	Big Blue®	Muta-genicity	Agree-ment[b]
Agaritine	Mouse	Oral	Lung				Lung	–	No
			Vascular system						
							Kidney	(+)	No
							Forestomach	(+)	No
							Liver	–	Yes
							Glandular stomach	–	Yes
4-Amino-biphenyl	Mouse	Oral	Blood vessels						
			Bladder	Bladder	+	Yes			
			Liver	Liver	+	Yes			
				Bone marrow	+	No			
2-Amino-3,4-dimethyl-imidazo[4,5-*f*]-quinoline (MeIQ)	Mouse	Oral	Caecum/colon				Colon	+	Yes
			Liver				Liver	+	Yes
			Forestomach				Forestomach	+	Yes
			Small intestine						

Table 13 (Contd)

Substance	Species	Route	Target organs in carcinogenicity studies	Organs examined in transgenic animal assays					
				Muta™-Mouse	Muta-genicity	Agree-ment[b]	Big Blue®	Muta-genicity	Agree-ment[b]
(contd)			Blood vessels						
							Bone marrow	+	No
							Heart	–	Yes
2-Amino-3,8-dimethyl-imidazo[4,5-*f*]-quinoxaline (MeIQx)	Rat	Oral	Liver				Liver	+	Yes
			Zymbal gland				Zymbal gland	+	Yes
			Clitoral gland						
			Skin						
							Colon	+	No
							Kidney	+	No
							Spleen	(+)	No
							Lung	–	Yes
							Testis	–	Yes
							Heart	–	Yes
							Brain	–	Yes
							Fat tissue	–	Yes

Table 13 (Contd)

Substance	Species	Route	Target organs in carcinogenicity studies	Organs examined in transgenic animal assays					
				Muta™-Mouse	Muta-genicity	Agree-ment[b]	Big Blue®	Muta-genicity	Agree-ment[b]
(contd)							Skeletal muscle	–	Yes
2-Amino-1-methyl-6-phenyl-imidazo[4,5-*b*]pyridine (PhIP)	Mouse	Oral	Blood (lymphoma)						
			Lung						
				Colon	+	No			
				Small intestine	+	No			
				Liver	(+)	No			
				Kidney	–	Yes			
	Rat	Oral	Colon				Colon	+	Yes
			Caecum				Caecum	+	Yes
			Small intestine						
			Mammary				Mammary	+	Yes
			Prostate				Prostate	+	Yes
							Kidney	+	No

Table 13 (Contd)

Substance	Species	Route	Target organs in carcinogenicity studies	Organs examined in transgenic animal assays					
				Muta™-Mouse	Muta-genicity	Agree-ment[b]	Big Blue®	Muta-genicity	Agree-ment[b]
2-Amino-3-methylimi-dazo[4,5-*f*]quinoline (IQ)	Rat	Oral	Zymbal gland						
			Colon				Colon	+	Yes
			Small intestine						
			Liver				Liver	+	Yes
			Skin						
			Clitoral gland						
							Kidney	+	No
Benzene	Mouse	Oral	Adrenal gland						
			Blood (no details)				Bone marrow	+	Yes
							Spleen	+	Yes
			Liver						
			Lung				Lung	–	No
			Ovary						
			Preputial gland						
			Zymbal gland						

Table 13 (Contd)

Substance	Species	Route	Target organs in carcinogenicity studies	Organs examined in transgenic animal assays					
				Muta™-Mouse	Muta-genicity	Agree-ment[b]	Big Blue®	Muta-genicity	Agree-ment[b]
(contd)			Mammary gland						
	Mouse	Inhal.	Blood (lymphoma)				Spleen	+	Yes
							Liver	−	Yes
			Blood (leukaemia)						
			Lung				Lung	+	Yes
			Zymbal gland						
Benzo[*a*]py-rene	Mouse	Oral	Blood (lymphoma)	Spleen	+	Yes			
			Blood (leukaemia)	Bone marrow	+	Yes			
			Lung	Lung	+	Yes			
			Stomach	Glandular stomach	+	Yes			
				Forestomach	+	Yes			
			Thymus						

Table 13 (Contd)

Substance	Species	Route	Target organs in carcinogenicity studies	Organs examined in transgenic animal assays					
				Muta™-Mouse	Muta-genicity	Agree-ment[b]	Big Blue®	Muta-genicity	Agree-ment[b]
(contd)				Heart	+	No			
				Ileum	+	No			
				Kidney	+	No			
				Liver	+	No			
				Colon	+	No			
				Mammary gland	+	No			
				Oral cavity	+	No			
				Breast	+	No			
				Tongue	+	No			
				Brain	–	Yes			
1,3-Butadiene	Mouse	Inhal.	Heart						
			Blood (lymphoma)						
				Bone marrow	–	Yes	Bone marrow	+	No
			Lung	Lung	+	Yes			
			Forestomach						

Table 13 (Contd)

Substance	Species	Route	Target organs in carcinogenicity studies	Organs examined in transgenic animal assays					
				Muta™-Mouse	Muta-genicity	Agree-ment[b]	Big Blue®	Muta-genicity	Agree-ment[b]
(contd)			Harderian gland						
			Liver	Liver	–	No			
			Mammary gland						
			Ovary						
			Preputial gland						
			Kidney						
Cyclophos-phamide	Mouse	Parent.	Mammary gland						
			Ovary						
			Lung				Lung	+	Yes
			Liver				Liver	+	Yes
			Testis				Testis	–	No
			Bladder				Bladder	+	Yes
			Local sarcoma						
				Bone marrow	+	No	Bone marrow	–	Yes

Table 13 (Contd)

Substance	Species	Route	Target organs in carcinogenicity studies	Organs examined in transgenic animal assays					
				Muta™-Mouse	Muta-genicity	Agree-ment[b]	Big Blue®	Muta-genicity	Agree-ment[b]
(contd)							Kidney	–	Yes
							Spleen	+	No
7,12-Dimethyl-benz[*a*]an-thracene (DMBA)	Mouse	Parent.	Local sarcoma						
			Lung						
			Blood (lymphoma)	Thymus	+	Yes			
			Liver	Liver	+	Yes			
				Bone marrow	+	No			
				Skin	+	No			
				Colon	+	No			
				Testis	+	No			
				Kidney	+	No			
Ethylene oxide	Mouse	Inhal.	Lung				Lung	+	Yes
			Harderian gland						
			Blood (lymphoma)				Spleen	–	No

Table 13 (Contd)

Substance	Species	Route	Target organs in carcinogenicity studies	Organs examined in transgenic animal assays					
				Muta™-Mouse	Muta-genicity	Agree-ment[b]	Big Blue®	Muta-genicity	Agree-ment[b]
(contd)			Uterus						
			Mammary gland						
							Bone marrow	–	Yes
							Germ cells	–	Yes
Ethyl methane-sulfonate	Mouse	Parent.	Lung						
			Kidney						
			Thymus						
				Bone marrow	+	No			
				Liver	+	No			
				Brain	–	Yes			
N-Ethyl-*N*-nitrosourea	Mouse	Parent.	Liver	Liver	+	Yes	Liver	+	Yes
			Harderian gland						
			Lympho-reticular	Spleen	+	Yes	Spleen	+	Yes
			Ovary						

Table 13 (Contd)

Substance	Species	Route	Target organs in carcinogenicity studies	Organs examined in transgenic animal assays					
				Muta™-Mouse	Muta-genicity	Agree-ment[b]	Big Blue®	Muta-genicity	Agree-ment[b]
(contd)			Mammary gland						
				Bone marrow	+	No			
				Bladder	+	No			
				Lung	+	No	Lung	+	No
				Kidney	+	No			
				Heart	+	No			
				Testis	+	No			
				Germ cells	+	No	Germ cells	+	No
				Brain	–	Yes			
Hydrazine & salts	Mouse	Oral	Liver	Liver	–	No			
			Lung	Lung	–	No			
			Blood (lymphoma)						
				Bone marrow	–	Yes			

Table 13 (Contd)

Substance	Species	Route	Target organs in carcinogenicity studies	Organs examined in transgenic animal assays					
				Muta™-Mouse	Muta-genicity	Agree-ment[b]	Big Blue®	Muta-genicity	Agree-ment[b]
N-Methyl-*N*-nitro-*N*-nitroso-guanidine (MNNG)	Mouse	Oral	Intestine						
			Stomach	Stomach	+	Yes			
				Bone marrow	–	Yes			
				Liver	–	Yes			
4-Nitroquino-line 1-oxide	Mouse	Parent.	Lung	Lung	+	Yes			
				Bone marrow	+	No			
				Liver	+	No			
				Spleen	–	Yes			
				Testis	–	Yes			
				Stomach	–	Yes			
				Kidney	–	Yes			
N-Nitroso-dimethyl-amine	Mouse	Oral	Lung				Lung	–	No
			Liver	Liver	+	Yes	Liver	+	Yes
			Kidney						
			Vascular system						

Table 13 (Contd)

Substance	Species	Route	Target organs in carcinogenicity studies	Organs examined in transgenic animal assays					
				Muta™-Mouse	Muta-genicity	Agree-ment[b]	Big Blue®	Muta-genicity	Agree-ment[b]
(contd)				Nasal mucosa	–	Yes			
							Bladder	–	Yes
							Forestomach	–	Yes
	Mouse	Parent.	Vascular system						
			Lung	Lung	–	No	Lung	+	Yes
				Liver	+	No	Liver	+	No
				Spleen	–	Yes			
				Kidney	–	Yes	Kidney	+	No
							Bladder	–	Yes
							Bone marrow	–	Yes
							Testis	–	Yes
N-Nitrosodi-*n*-propylamine	Mouse	Parent.	Nasal cavity						
			Intestine						
			Liver	Liver	+	Yes			
				Lung	+	No			

Table 13 (Contd)

Substance	Species	Route	Target organs in carcinogenicity studies	Organs examined in transgenic animal assays					
				Muta™-Mouse	Muta-genicity	Agree-ment[b]	Big Blue®	Muta-genicity	Agree-ment[b]
(contd)				Kidney	+	No			
				Bone marrow	+	No			
				Bladder	–	Yes			
				Testis	–	Yes			
Procarbazine	Mouse	Parent.	Lung	Lung	+	Yes			
			Blood (lymphoma)	Spleen	+	Yes			
			Blood (leukaemia)	Bone marrow	+	Yes			
			Kidney	Kidney	+	Yes			
			Uterus						
				Liver	+	No			
				Testis	+	No			
				Brain	–	Yes			
Quinoline	Mouse	Parent.	Liver	Liver	+	Yes			
				Bone marrow	–	Yes			
				Spleen	–	Yes			

Table 13 (Contd)

Substance	Species	Route	Target organs in carcinogenicity studies	Organs examined in transgenic animal assays					
				Muta™-Mouse	Muta-genicity	Agree-ment[b]	Big Blue®	Muta-genicity	Agree-ment[b]
(contd)				Lung	–	Yes			
				Kidney	–	Yes			
				Testis	–	Yes			
Tris(2,3-di-bromo-propyl)phos-phate	Mouse	Oral	Lung						
			Kidney				Kidney	+	Yes
			Forestomach				Stomach	–	No
			Liver				Liver	–	No
Urethane (ethyl carbamate)	Mouse	Oral	Blood (lymphoma)						
			Blood (leukaemia)						
			Liver				Liver	+	Yes
			Lung				Lung	+	Yes
			Skin/subcuta-neous tissue						
			Thymus						

Table 13 (Contd)

Substance	Species	Route	Target organs in carcinogenicity studies	Organs examined in transgenic animal assays					
				Muta™-Mouse	Muta-genicity	Agree-ment[b]	Big Blue®	Muta-genicity	Agree-ment[b]
(contd)			Harderian gland						
							Forestomach	+	No
	Mouse	Parent.	Lung	Lung	+	Yes			
			Liver	Liver	+	Yes			
				Spleen	+	No			
				Bone marrow	+	No			

+: increased mutagenic activity in transgenic animals; (+): weak positive results in transgenic animals; −: no increase in mutagenic activity in transgenic animals; Inhal.: Inhalation; Parent.: Parenteral

[a] For references, see Master Table (Appendix 1). Differences in strain and gender were not taken into account.
[b] Agreement between transgenic animal assay and carcinogenicity study concerning results on target organs.

- For compounds that have multiple target organs in carcinogenesis studies, such as 4-aminobiphenyl, 2-amino-3,4-dimethylimidazo[4,5-*f*]quinoline, 2-amino-3,4-dimethylimidazo[4,5-*f*]quinoxaline, 2-amino-1-methyl-6-phenylimidazo[4,5-*b*]pyridine, 2-amino-3-methylimidazo[4,5-*f*]quinoline, benzene, benzo[*a*]pyrene, 1,3-butadiene, cyclophosphamide, ethylene oxide, 7,12-dimethylbenz[*a*]anthracene, *N*-ethyl-*N*-nitrosourea, *N*-nitrosodimethylamine, procarbazine and urethane, mutations were found in most of the target organs. In some single target organs, no mutations have been detected (e.g. in the lung for benzene, the liver for butadiene, the testis for cyclophosphamide).

- For some compounds, numerous organs have been investigated in the transgenic assays, which also included organs where no tumours had occurred in the carcinogenicity studies (e.g. agaritine, 2-amino-3,4-dimethylimidazo[4,5-*f*]quinoline, 2-amino-3,4-dimethylimidazo[4,5-*f*]quinoxaline, 2-amino-1-methyl-6-phenylimidazo[4,5-*b*]pyridine, benzo[*a*]pyrene, 7,12-dimethylbenz-[*a*]anthracene, *N*-ethyl-*N*-nitrosourea, 4-nitroquinoline 1-oxide). For benzo[*a*]pyrene and 7,12-dimethylbenz[*a*]anthracene, all organs investigated in the transgenic animal mutagenicity assays were positive. Similarly for *N*-ethyl-*N*-nitrosourea, nearly all organs were positive. Although this occurred for several compounds, it cannot be explained by insufficient target organ specificity. Instead, it leads to the conclusion that although genotoxicity is expressed in nearly all organs in the body, tumours do not develop in all these organs, probably due to factors other than genotoxicity.

These results suggest that the transgenic animal mutagenicity assay is useful, in part, for the prediction of target organs for carcinogenesis. However, a positive response in a specific tissue does not necessarily mean that tumours will be induced by that chemical in that specific organ.

10.1.2 Analysis of the predictivity for the liver as target organ

In contrast to other target organs, the liver has been analysed in most of the transgenic animal mutagenicity assays. Therefore, a comparison of the mutagenicity in this target organ with the outcome of carcinogenicity studies could be made for a larger number of compounds than was possible for the pattern of target organs (Table 13). Such an analysis is summarized in Table 14.

Concerning the target organ liver, for most substances in Table 14, there is agreement between the results in the carcinogenicity studies and the transgenic animal mutagenicity assays. However, for some compounds that are known hepatocarcinogens, the transgenic animal assays were negative.

Most of the compounds with liver as target organ in carcinogenicity studies, but negative outcome in the transgenic animal mutagenicity assays, are compounds for which generally a non-genotoxic mode of action of carcinogenicity is assumed (e.g. chloroform, di-(2-ethylhexyl) phthalate, 2,3,7,8-tetrachlorodibenzo-*p*-dioxin and tetrachloromethane). A further analysis of these compounds is presented in section 10.2.1 (see also Table 17 below).

1,3-Butadiene, heptachlor and hydrazine induced liver tumours in carcinogenicity studies but did not increase mutagenic activity in transgenic animal assays, although genotoxic mechanisms of carcinogenicity are suggested. For 1,3-butadiene, positive results were obtained in the same experiment on the Muta™Mouse in the target organ lung, indicating organ-specific differences in sensitivity. The heptachlor study is limited by the fact that a small number of animals were used and high interindividual differences were observed (see Table 18 below). Therefore, at this time, the results of this study are not suitable for suggesting non-genotoxic mechanisms of liver carcinogenesis (mechanisms still under discussion). The validity of the transgenic animal mutagenicity assay with hydrazine (see section 6.2) is sufficient, but other data on genotoxicity indicated that repeated instead of single exposure is necessary for detection of mutagenic effects with this weak mutagen.

Aflatoxin B1 is known to induce no or limited tumorigenic effects in mouse liver, but strong tumorigenic effect in rats. This species specificity is linked to differences in metabolic activation of aflatoxin in the two species. Aflatoxin was evaluated in both Big Blue® mice and rats, providing the opportunity to evaluate the ability of the transgenic animal models to identify this species specificity. Aflatoxin B1 induced liver tumours in F344 rats after gavage

Table 14. Comparison of carcinogenic effects in the liver with outcome of the transgenic animal mutagenicity assays in the liver[a]

Substance	Route	Species	Target organ in carcinogenicity	Mutagenic in Muta™Mouse	Agreement with carcinogenicity	Mutagenic in Big Blue®	Agreement with carcinogenicity
2-Acetylaminofluorene	Oral	Mouse	+	+	Yes	+	Yes
Aflatoxin B1	Parenteral	Mouse	+ only newborn			–	No
	Oral	Rat	+			+	Yes
Agaritine	Oral	Mouse	–			–	Yes
4-Aminobiphenyl	Oral	Mouse	+	+	Yes		
2-Amino-3,4-dimethyl-imidazo[4,5-*f*]quinoline	Oral	Mouse	+			+	Yes
2-Amino-3,8-dimethyl-imidazo[4,5-*f*]quinox-aline	Oral	Mouse and rat	+			+	Yes
2-Amino-1-methyl-6-phenylimidazo[4,5-*b*]pyridine	Oral	Mouse	–	(+)	(No)		
2-Amino-3-methylimid-azo[4,5-*f*]quinoline (IQ)	Oral	Mouse	+	+	Yes		
	Oral	Rat	+			+	Yes
ortho-Anisidine	Oral	Mouse	–			–	Yes
Benzene	Inhalation	Mouse	–			–	Yes

Table 14 (Contd)

Substance	Route	Species	Target organ in carcinogenicity	Mutagenic in Muta™Mouse	Agreement with carcinogenicity	Mutagenic in Big Blue®	Agreement with carcinogenicity
Benzo[*a*]pyrene	Oral	Mouse	–	+	No		
	Parenteral	Mouse	+	+	Yes	+	Yes
1,3-Butadiene	Inhalation	Mouse	+	–	No		
Chlorambucil	Parenteral	Mouse	–	+	No		
Chloroform	Inhalation	Mouse	+			–	No
Cyclophosphamide	Parenteral	Mouse	+			+	Yes
2,4-Diaminotoluene	Oral	Mouse	+			+	Yes
2,6-Diaminotoluene	Oral	Mouse	–			–	Yes
Di-(2-ethylhexyl) phthalate (DEHP)	Oral	Mouse	+			–	No
5-(*p*-Dimethylamino-phenylazo)-benzothia-zole	Oral	Rat	–			+	No
6-(*p*-Dimethylamino-phenylazo)-benzothia-zole	Oral	Rat	+			+	Yes
7,12-Dimethyl-benz[*a*]anthracene	Parenteral	Mouse	+	+	Yes		
Ethyl methanesulfonate	Parenteral	Mouse	–	+	No		

Table 14 (Contd)

Substance	Route	Species	Target organ in carcinogenicity	Mutagenic in Muta™Mouse	Agreement with carcinogenicity	Mutagenic in Big Blue®	Agreement with carcinogenicity
N-Ethyl-*N*-nitrosourea	Parenteral	Mouse	+	+	Yes	+	Yes
Heptachlor	Oral	Mouse	+			–	No
Hydrazine and salts	Oral	Mouse	+	–	No		
(+)-Limonene	Oral	Rat	–			–	Yes
N-Methyl-*N*-nitrosourea	Parenteral	Mouse	+			+	Yes
N-Methyl-*N'*-nitro-*N*-nitrosoguanidine	Oral	Mouse	–	–	Yes		
Mitomycin C	Parenteral	Mouse	–	–	Yes		
4-Nitroquinoline 1-oxide	Parenteral	Mouse	–	+	No		
N-Nitrosodiethylamine	Parenteral	Mouse	+	+	Yes		
N-Nitrosodimethylamine	Oral	Mouse	+	+	Yes	+	Yes
	Oral	Rat	+			+	Yes
	Parenteral	Mouse	–	+	No	±	Yes
N-Nitrosodi-*n*-propylamine	Parenteral	Mouse	+	+	Yes		
Phenobarbital	Oral	Mouse	+	–	No	(+)	Yes
Procarbazine	Parenteral	Mouse	–	+	No		
Quinoline	Parenteral	Mouse	+	+	Yes		

Table 14 (Contd)

Substance	Route	Species	Target organ in carcinogenicity	Mutagenic in Muta™Mouse	Agreement with carcinogenicity	Mutagenic in Big Blue®	Agreement with carcinogenicity
Sodium saccharin	Oral	Rat	–			–	Yes
2,3,7,8-Tetrachlorodi-benzo-*p*-dioxin	Oral	Rat	+			–	No
Tetrachloromethane (carbon tetrachloride)	Oral	Mouse	+	–	No		
Tris(2,3-dibromopropyl)-phosphate	Oral	Mouse	+			–	No
Urethane (ethyl carbamate)	Oral	Mouse	+			+	Yes
	Parenteral	Mouse	+	+	Yes		

(+): weak positive result in transgenic animal assay; +: liver is a target organ concerning the specified end-point; –: liver is not a target organ concerning the specified end-point.

[a] Differences in strain and gender were not taken into account.

(IARC, 1993a). In F344 Big Blue® rats, increased mutation/mutant frequency was observed in the liver after gavage (Davies et al., 1997) or single intraperitoneal administration (Dycaico et al., 1996). Dycaico et al. (1996) also studied the effects of aflatoxin B1 in Big Blue® mice in parallel experiments. Even after an intraperitoneal administration at a dose 10-fold higher than the dose used in rats (2.5 versus 0.25 mg/kg of body weight), no significant increase in mutant frequency was detected, indicating a higher sensitivity of rats to this carcinogen. However, after repeated oral application of higher doses (8 mg/kg of body weight), a weak significant increase in mutation frequency (less than 2-fold) was detected in Big Blue® mouse liver (Autrup et al., 1996). Liver tumours were never observed in carcinogenicity studies in mice except in newborn mice after intraperitoneal applications. This carcinogenic effect in newborn mice might be due to a different balance in activation and detoxification in newborn mice compared with adult mice and/or to proliferation of liver cells in newborn mice. There are some studies where negative results were obtained for the liver in carcinogenicity studies but positive results in the transgenic animal mutagenicity assay (2-amino-1-methyl-6-phenylimidazo[4,5-*b*]pyridine, chlorambucil, 5-(*p*-dimethylaminophenylazo)-benzothiazole, ethyl methanesulfonate, 4-nitroquinoline 1-oxide and procarbazine). This is the same finding as reported above when analysing the pattern of target organs and again supports the idea that processes other than genotoxicity are involved in carcinogenicity.

For several substances, the negative outcome in carcinogenicity studies was confirmed in transgenic animal mutagenicity assays: mutagenic activity was absent in the liver in agaritine, *o*-anisidine, benzene, 2,6-diaminotoluene, (+)-limonene, *N*-methyl-*N*′-nitro-*N*-nitrosoguanidine, mitomycin C and sodium saccharin.

10.1.3 Conclusion

In transgenic animal assays, mutagenic events can be detected in any organ. Thus, target organs of carcinogenicity may be predicted in this test system. The available data have shown that in most cases, mutations have been detected in the target organs of the carcinogenicity studies. On the other hand, carcinogens with presumed non-genotoxic mechanisms of carcinogenicity are generally not mutagenic in the transgenic animal assays.

For several compounds, numerous organs were investigated in the transgenic animal mutagenicity assay, including target and non-target organs in carcinogenicity studies. It could be shown that many organs investigated in the transgenic animal mutagenicity assays were positive that were not target organs in carcinogenicity studies. This cannot be explained by insufficient specificity with regard to target organs for carcinogenicity, but leads to the conclusion that genotoxicity is expressed in several organs in the body and that tumours do not develop in all these organs, probably due to other factors.

10.2 Comparison of results of carcinogenicity studies with results from transgenic animal mutagenicity assays

One of the crucial questions for the use and acceptance of transgenic animal systems in toxicological testing is their ability to predict carcinogenic effects. In a first approach, researchers tried to validate the system for positive predictivity. Therefore, predominantly carcinogenic substances have been investigated in transgenic animal mutagenicity assays. Few data are available on non-genotoxic carcinogens and non-carcinogenic compounds.

Forty-seven compounds have been tested in transgenic mice and in carcinogenicity studies on this species (see Table 15), and 10 compounds have been tested in studies on these two end-points using rats (Table 16).

For the majority of the substances evaluated in this document, there is good agreement between the results in the carcinogenicity studies and the transgenic assays for any tissue. Most of the numerous positive substances with respect to carcinogenicity gave positive results in the transgenic assays in mice or rats. The Muta™Mouse system agreed with carcinogenicity for 25 of 32 substances, whereas the Big Blue® mouse system agreed for 19 of 24 substances (agaritine not included because of inconclusive results). Ten substances were tested by both transgenic mouse systems and carcinogenicity tests as well, and eight substances gave positive results in all three test systems. The results of studies with the Big Blue® rat showed agreement for 6 of 10 compounds tested (Table 16).

Table 15. Predictivity of carcinogenic effects using results in transgenic mouse assays

Substance	Carcinogenicity			Transgenic Muta™Mouse assays		Transgenic Big Blue® mouse assays	
	Classification[a]		Results in carcinogenicity studies on mice	Results	Agreement with carcinogenicity study results	Results	Agreement with carcinogenicity study results
	IARC	MAK					
2-Acetylamino-fluorene	No eval.	No eval.	Positive	Positive	Yes	Positive	Yes
Acrylamide	2A	K2	Positive	Positive	Yes	No data	Not applicable
Aflatoxin B1	1	No eval.	Positive	No data	Not applicable	Weak positive	Yes
Agaritine	3	No eval.	Negative Limited evidence for a fungal derivative	No data	Not applicable	Weak positive for mushroom extracts	Yes
4-Aminobiphenyl	1	K1	Positive	Positive	Yes	No data	Not applicable
2-Amino-3,4-dimethylimidazo-[4,5-*f*]quinoline	2B	No eval.	Positive	No data	Not applicable	Positive	Yes
2-Amino-3,8-dimethylimidazo-[4,5-*f*]quinox-aline	2B	No eval.	Positive	No data	Not applicable	Positive	Yes

Table 15 (Contd)

Substance	Carcinogenicity			Transgenic Muta™Mouse assays		Transgenic Big Blue® mouse assays	
	Classification[a]		Results in carcinogenicity studies on mice	Results	Agreement with carcinogenicity study results	Results	Agreement with carcinogenicity study results
	IARC	MAK					
2-Amino-1-methyl-6-phenylimidazo[4,5-*b*]-pyridine	2B	No eval.	Positive	Positive	Yes	No data	Not applicable
2-Amino-3-methylimidazo[4,5-*f*]quinoline (IQ)	2A	No eval.	Positive	No data	Not applicable	Positive	Yes
ortho-Anisidine	2B	No eval.	Positive	No data	Not applicable	Positive	Yes
Asbestos crocidolite	1	K1	Positive	No data	Not applicable	Positive	Yes
Benzene	1	K1	Positive	No data	Not applicable	Positive	Yes
Benzo[*a*]pyrene	2A	K2	Positive	Positive	Yes	Positive	Yes
Bromomethane	3	K3B	Negative	Negative	Yes	No data	Not applicable
1,3-Butadiene	2A	K1	Positive	Positive	Yes	Positive	Yes
Chlorambucil	1	No eval.	Positive	Positive	Yes	No data	Not applicable

Table 15 (Contd)

Substance	Carcinogenicity			Transgenic Muta™Mouse assays		Transgenic Big Blue® mouse assays	
	Classification[a]		Results in carcinogenicity studies on mice	Results	Agreement with carcinogenicity study results	Results	Agreement with carcinogenicity study results
	IARC	MAK					
Chloroform	2B	K4 non-genotoxic	Positive, non-genotoxic	No data	Not applicable	Negative	No (special mechanisms)
Cyclophos-phamide	1	No eval.	Positive	Positive	Yes	Positive	Yes
2,4-Diamino-toluene	2B	K2	Positive	No data	Not applicable	Positive	Yes
2,6-Diamino-toluene	No eval.	No eval.	Negative	No data	Not applicable	Negative	Yes
1,2-Dibromo-ethane	2A	K2	Positive	Positive	Yes	No data	Not applicable
1,2-Dibromo-3-chloropropane	2B	K2	Positive	Weak positive	Yes	No data	Not applicable
1,2-Dichloro-ethane	2B	K2	Positive	Negative	No	No data	Not applicable
Di-(2-ethylhexyl) phthalate (DEHP)	2B	K4 non-genotoxic	Positive, non-genotoxic	No data	Not applicable	Negative	No (special mechanisms)

Table 15 (Contd)

Substance	Carcinogenicity			Transgenic Muta™Mouse assays		Transgenic Big Blue® mouse assays	
	Classification[a]		Results in carcinogenicity studies on mice	Results	Agreement with carcinogenicity study results	Results	Agreement with carcinogenicity study results
	IARC	MAK					
7,12-Dimethyl-benz[*a*]anthra-cene	No eval.	No eval.	Positive	Positive	Yes	No data	Not applicable
Ethylene oxide	1	K2	Positive	No data	Not applicable	Positive	Yes
Ethyl methane-sulfonate	2B	No eval.	Positive	Positive	Yes	No data	Not applicable
N-Ethyl-*N*-nitrosourea	2A	No eval.	Positive	Positive	Yes	Positive	Yes
Heptachlor	2B	K3B	Positive, probably non-genotoxic	No data	Not applicable	Negative	No (special mechanisms)
Hydrazine or hydrazine sulfate	2B	K2	Positive	Negative	No	No data	Not applicable
Methyl methane-sulfonate	2B	No eval.	Positive	Weak positive	Yes	Negative	No
N-Methyl-*N'*-nitro-*N*-nitroso-guanidine	2A	No eval.	Positive	Positive	Yes	No data	Not applicable

Table 15 (Contd)

Substance	Carcinogenicity			Transgenic Muta™Mouse assays		Transgenic Big Blue® mouse assays	
	Classification[a]		Results in carcinogenicity studies on mice	Results	Agreement with carcinogenicity study results	Results	Agreement with carcinogenicity study results
	IARC	MAK					
4-(Methylnitrosamino)-1-(3-pyridyl)-1-butanone	2B	No eval.	Positive	Positive	Yes	No data	Not applicable
N-Methyl-*N*-nitrosourea	2A	No eval.	Positive	Positive	Yes	Positive	Yes
Mitomycin C	2B	No eval.	Positive	Negative (but clastogenic)	No	No data	Not applicable
4-Nitroquinoline 1-oxide	No eval.	No eval.	Positive	Positive	Yes	No data	Not applicable
N-Nitroso-diethylamine	2A	K2	Positive	Positive	Yes	No data	Not applicable
N-Nitroso-dimethylamine	2A	K2	Positive	Positive	Yes	Positive	Yes
N-Nitrosodi-*n*-propylamine	2B	K2	Positive	Positive	Yes	No data	Not applicable

Table 15 (Contd)

Substance	Carcinogenicity			Transgenic Muta™Mouse assays		Transgenic Big Blue® mouse assays	
	Classification[a]		Results in carcinogenicity studies on mice	Results	Agreement with carcinogenicity study results	Results	Agreement with carcinogenicity study results
	IARC	MAK					
Phenobarbital	2B	No eval.	Positive, probably non-genotoxic	Negative	No	Negative[b]	Yes
Procarbazine	2A	No eval.	Positive	Positive	Yes	No data	Not applicable
β-Propiolactone	2B	K2	Positive	Positive	Yes	No data	Not applicable
N-Propyl-*N*-nitrosourea	No eval.	No eval.	No data (positive in rats)	Positive	Not applicable	No data	Not applicable
Quinoline	2A	No eval.	Positive	Positive	Yes	No data	Not applicable
Tetrachloro-methane	2B	K4 compen-satory cell regenera-tion	Positive, non-genotoxic	Negative	No (special mechanisms)	No data	Not applicable
Trichloro-ethylene	2A	K1	Positive	Negative	No	No data	Not applicable

Table 15 (Contd)

Substance	Carcinogenicity			Transgenic Muta™Mouse assays		Transgenic Big Blue® mouse assays	
	Classification[a]		Results in carcinogenicity studies on mice	Results	Agreement with carcinogenicity study results	Results	Agreement with carcinogenicity study results
	IARC	MAK					
Tris(2,3-di-bromopropyl)-phosphate	2A	No eval.	Positive	No data	Not applicable	Positive	Yes
Urethane	2B	K2	Positive	Positive	Yes	Positive	Yes

[a] MAK categories: MAK K1, carcinogenic to humans; MAK K2, probably carcinogenic to humans, sufficient evidence for carcinogenicity in animals; MAK K3B, limited evidence for carcinogenicity from in vitro or in vivo studies but not sufficient for a classification to any other group; MAK K4, substance with carcinogenic effects, but genotoxicity does not play a decisive role; No eval.: No evaluation. For IARC categories, see footnote b of Table 2.

[b] After clonal correction (Thybaud et al., 2003).

Table 16. Predictivity of carcinogenic effects using results in transgenic rat assays

Substance	Carcinogenicity			Transgenic Big Blue® rat assays	
	Classification[a]		Results in carcinogenicity studies on rats	Results	Agreement with carcinogenicity study results
	IARC	MAK			
2-Amino-3-methyl-imidazo[4,5-*f*]quinoline (MeIQx)	Group 2B	No evaluation	Positive	Positive	Yes
2-Amino-1-methyl-6-phenylimidazo[4,5-*b*]pyridine (PhIP)	Group 2B	No evaluation	Positive	Positive	Yes
2-Amino-3-methyl-imidazo[4,5-*f*] quinoline (IQ)	Group 2A	No evaluation	Positive	Positive	Yes
5-(*p*-Dimethyl-aminophenylazo)-benzothiazole	No evaluation	No evaluation	No increased tumour incidence compared with the analogue 6BT	Positive	No
6-(*p*-Dimethyl-aminophenylazo)-benzothiazole	No evaluation	No evaluation	Positive	Positive	Yes
(+)-Limonene	Group 3	No evaluation	Positive, alpha-2u-globulin involved	Negative	No (special mechanisms)
N-Nitrosodimethyl-amine	Group 2A	K2	Positive	Positive	Yes

Table 16 (Contd)

Substance	Carcinogenicity			Transgenic Big Blue® rat assays	
	Classification[a]		Results in carcinogenicity studies on rats	Results	Agreement with carcinogenicity study results
	IARC	MAK			
Sodium saccharin	Group 3	No evaluation	Positive	Negative	No (special mechanisms)
Tamoxifen	Group 1	No evaluation	Positive	Positive	Yes
2,3,7,8-Tetrachloro-dibenzo-*p*-dioxin (TCDD)	Group 1	K4	Positive	Negative	No (special mechanisms)

6BT = 6-(*p*-dimethylaminophenylazo)-benzothiazole

[a] MAK categories: MAK K2, probably carcinogenic to humans, sufficient evidence for carcinogenicity in animals; MAK K4, substance with carcinogenic effects, but genotoxicity does not play a decisive role. For IARC classifications, see footnote b to Table 2.

Two compounds (2,6-diaminotoluene and bromomethane) are the only non-carcinogens that have been assessed in transgenic mouse systems in this document. Both compounds scored negative in these gene mutation tests with transgenic animals.

Gene mutation assays with transgenic animals can explain and/or confirm (conflicting) results from carcinogenicity studies. For *o*-anisidine, conflicting results have been obtained with other genotoxicity assays. The transgenic mutation assay now proves the supposed genotoxic mode of action in the target organ bladder. For asbestos, the majority of genotoxity studies were negative. However, the transgenic assay revealed genotoxicity in the target organ lung. These results indicate that the transgenic assays can give good evidence for the genotoxic mode of action of carcinogens.

There is one compound, 5-(*p*-dimethylaminophenylazo)-benzothiazole, that gave a negative result in the carcinogenicity assays but a positive result in the transgenic rat assay. However, because the MTD was not reached in the carcinogenicity study, the results of the transgenic assay indicate that 5-(*p*-dimethylaminophenylazo)-benzothiazole is a putative carcinogen. This impression is enforced by results with the structural analogue 6-(*p*-dimethylaminophenylazo)-benzothiazole, which demonstrated similar mutagenic activity in parallel experiments (Fletcher et al., 1999).

There are several compounds with positive results in the carcinogenicity studies and negative results in the transgenic assays. There may be several reasons for this disagreement:

1) The compound is predominantly a clastogen: This may hold for hydrazine sulfate, mitomycin C and trichloroethylene, where clear positive results have been obtained in the micronucleus test, but equivocal results in some other genotoxicity assays.
2) The compound is a non-genotoxic carcinogen: These compounds are discussed below.
3) The study design of the transgenic assay was not optimal: For example, the negative result with hydrazine may be due to the fact that the compound was administered only once. Another compound where an inappropriate test design may be responsible for negative results is 1,2-dichloroethane. The study

design can also be questioned for several other compounds (see Table 18 below).

10.2.1 Non-genotoxic carcinogens

Some carcinogens are considered as non-genotoxic carcinogens if they show negative results in in vitro and in vivo genotoxicity studies. However, often there remains some doubt, as the target organ that showed tumours in the carcinogenicity study has not been examined in the standard mutagenicity tests (e.g. the mouse micronucleus test possibly does not detect mutagenic effects of a substance inducing liver tumours). Therefore, transgenic animal test systems, where any target organ can be examined, may be more suitable tools for investigating the mechanisms by which such carcinogens act.

There are several substances that are thought to cause carcinogenesis by non-genotoxic mechanisms or where the contribution of genotoxicity to the effects observed is not clear. Examples are chloroform, di-(2-ethylhexyl) phthalate, heptachlor, (+)-limonene, phenobarbital, sodium saccharin, 2,3,7,8-tetrachlorodibenzo-*p*-dioxin and tetrachloromethane. For all these compounds, negative results in the transgenic mutation assays have been obtained, supporting the proposed mode of action. For 2,3,7,8-tetrachlorodibenzo-*p*-dioxin, several target organs have been investigated, which were all negative. Therefore, the transgenic assays seem to be suitable tools for supporting non-genotoxic mechanisms of carcinogenicity (Table 17). On the other hand, there are some compounds for which genotoxicity is assumed, but not clearly detectable in standard *in vitro* genotoxicity assays (e.g. for the aromatic amines or for asbestos). Here, transgenic assays may also help to elucidate the mode of action.

10.2.2 Validity of data on transgenic animal mutation assays

The various studies using transgenic animal mutation assays have used varying protocols, which makes a comparison of the results more difficult. Therefore, a pragmatic approach was used. No examination of the validity was performed if any result obtained in at least one target organ of at least one study on transgenic animals was positive for the particular test substance. However, if all available data on transgenic animals gave negative results for a substance, the validity was checked (Table 18) using the validity criteria described in section 6.2. For the analysis, it is considered that a negative result using a robust protocol should be accepted as valid (Thybaud et al., 2003).

Table 17. Summary table for non-genotoxic compounds[a]

Compound	Tumour target organ	Presumed mode of action	Result of the transgenic animal assay in the target organ
Chloroform	Liver	Regenerative cell proliferation	Negative
Di-(2-ethylhexyl) phthalate	Liver	Interaction with regulatory processes for cell proliferation	Negative
Heptachlor	Liver	Inhibition of intracellular gap junction communication	Negative
(+)-Limonene	Kidney	Alpha-2u-globulin	Negative
Phenobarbital	Liver	Promoter	Negative
Sodium saccharin	Bladder	Mechanical irritation from precipitates	Negative
Tetrachloromethane	Liver	Compensatory cell regeneration	Negative
2,3,7,8-Tetrachlorodibenzo-*p*-dioxin	Liver Lung Thyroid Hard palate Nasal turbinates Tongue	Promoter	Negative

[a] For a number of compounds in the table, the studies were of limited validity (compare with Table 18).

10.2.3 *Evaluation of the predictivity for carcinogenicity in mice*

The sensitivity, specificity and predictive value of the Muta™Mouse assay and the Big Blue® mouse assays for carcinogenicity are documented in Table 19. Forty-six substances with data on carcinogenicity in mice and mutagenic effects in transgenic mice were available for the evaluation (see Table 15). Agaritine, which shows inconclusive results, and *N*-propyl-*N*-nitrosourea, which has not been tested for carcinogenicity in mice, were not included in this

Table 18. Validity of negative transgenic animal mutagenicity studies[a]

Substance	Species/strain, number per dose group	Dose regimen	MTD reached	Post-treatment time // no. of plaques per organ examined // statistics	Organs examined	Validity	References
Bromo-methane	Muta™Mouse n = 2–4* males	Gavage, once, 0, 12.5 or 50 mg/kg bw 10 daily doses of 25 mg/kg bw	Questionable* No data on toxicity	14 d after the last treatment* (but sufficient for the stomach) // >148 000 // yes	Liver Lung Spleen Bone marrow Liver Stomach	Limited The same dose resulted in methyl-ation of the DNA in the same animals; mutant frequency not increased by this level of premutagenic lesions; spleen and bone marrow studied only at the low dose level	Pletsa et al. (1999)

Table 18 (Contd)

Substance	Species/strain, number per dose group	Dose regimen	MTD reached	Post-treatment time // no. of plaques per organ examined // statistics	Organs examined	Validity	References
Chloroform	Big Blue® mouse/B6C3F1 *n* = 10 females	Inhalation; 0, 50, 149, or 446 mg/m^3 6 h/d, 7 d/week; exposure duration 10, 30, 90, 180 days	Yes; bw decreased, relative liver weight increased, no deaths	10 d // >200 000 // yes	Liver	Sufficient Hepatocyte necrosis and karyomegaly and regenerative cell proliferation detected	Butterworth et al. (1998)
2,6-Diamino-toluene	Big Blue® mouse/B6C3F1 *n* = 5 males	0 or 1000* mg/kg in the diet for 30 or 90 d	Questionable* No deaths, no further data	1 d // >100 000* // yes	Liver	Limited Valid positive control	Hayward et al. (1995); Cunningham et al. (1996)
1,2-Dichloro-ethane	Muta™Mouse *n* = 1–3* males	Gavage, once, 0, 75 or 150 mg/kg bw	Questionable* No data on toxicity	7, 14 or 28 d // >93 000 // yes	Liver	Limited Valid positive control	Hachiya & Motohashi (2000)

Table 18 (Contd)

Substance	Species/strain, number per dose group	Dose regimen	MTD reached	Post-treatment time // no. of plaques per organ examined // statistics	Organs examined	Validity	References
Di-(2-ethyl-hexyl) phthalate	Big Blue® mouse/ C57BL/6 *n* = 3* females	O, 3000 or 6000 mg/kg in the diet for 119 d	Questionable* No effects on bw, no further data on toxicity	1 d // ~180 000 for control and ~100 000* for treated mice // yes	Liver	Limited Valid positive control	Gunz et al. (1993)
Heptachlor	Big Blue® mouse/ C57BL/6 *n* = 3* females	0, 10 or 20 mg/kg in the diet for 119 d	Questionable* No effects on bw, no further data on toxicity	1 d // ~180 000 for control and ~100 000* for treated mice // yes	Liver	Limited Valid positive control; marked animal-to-animal variation with heptachlor	Gunz et al. (1993)
Hydrazine	Muta™Mouse *n* = 5 males	Gavage, once, 0, 135, 270, 350 or 400 mg/kg bw	Yes Lethal effects at 400 mg/kg bw	14 d or 56 d // >200 000 // yes	Liver Lung Bone marrow	Sufficient But repeated exposure might induce genotoxicity in weak mutagens	Douglas et al. (1995b)

Table 18 (Contd)

Substance	Species/strain, number per dose group	Dose regimen	MTD reached	Post-treatment time // no. of plaques per organ examined // statistics	Organs examined	Validity	References
(+)-Limonene	Rat, F344 *n* = 6–7 males	0 or 360–680* mg/kg bw per day via the diet for 10 d	Questionable* No data on toxicity	14 d* // >125 000 (*n* = 7) // yes	Liver Kidney	Limited One dose, gavage, used in carcinogenicity studies, valid positive control	Turner et al. (2001)
Mitomycin C	Muta™Mouse *n* = 1–3* males	1 × i.p., 0, 1 or 2 mg/kg bw or daily i.p., 1 or 2 mg/kg bw for 5 d	Questionable* Cytotoxicity in bone marrow at 2 mg/kg bw, no further data	Bone marrow 7–21 d, liver 7–8 d* // >500 000 except single dose, bone marrow, 21 d post-treatment time (77 000 and 154 000) // no data*	Bone marrow Liver	Limited In the same mice, micronucleus induction in the bone marrow; valid bone marrow positive control; no concurrent control after repeated exposure	Suzuki et al. (1993)

Table 18 (Contd)

Substance	Species/strain, number per dose group	Dose regimen	MTD reached	Post-treatment time // no. of plaques per organ examined // statistics	Organs examined	Validity	References
Sodium saccharin	Rat, F344 *n* = 6–7 males	0 or 5%* (~2500 mg/kg bw) in the diet for 10 d*	No* Rats appeared healthy, no further data on toxicity	14 d* // >122 000 (*n* = 7) for the liver but only 40 450 for the bladder* // yes	Liver Bladder	Limited Dose used induced bladder tumours in long-term (2 years) studies, valid positive control	Turner et al. (2001)
TCDD	Rat, F344 *n* = 5 males and 5 females	0 or 2 µg/kg bw* twice weekly via gavage for 6 weeks	Questionable* Thymus weight decreased, liver weight increased, no further data	14 d // ~1 100 000, no data on control* // yes	Liver	Limited One dose, no positive control (also no historical data)	Thornton et al. (2001)
Tetra-chloro-methane	Muta™Mouse *n* = 5 males	0 or 80 mg/kg bw* once via gavage	Questionable* No effect on liver weight but liver cell regeneration, no further data	14 d* // >200 000 // yes	Liver	Limited One dose, valid positive control	Tombolan et al. (1999a)

Table 18 (Contd)

Substance	Species/strain, number per dose group	Dose regimen	MTD reached	Post-treatment time // no. of plaques per organ examined // statistics	Organs examined	Validity	References
Trichloro-ethylene	Muta™Mouse n = 5–10 females or n = 8–9 males	Inhalation of 0, 1090, 6190 or 16 890 mg/m^3 for 12 d	Questionable* No data given on toxicity	60 d (also 14 d in a further test, lung examined) // >200 000 // no data	Lung Liver Bone marrow Spleen Kidney Testis	Limited No positive (historical) control; in tissues with high cell turnover (e.g. bone marrow), mutations might be missed	Douglas et al. (1999)

*: limitations of the validity; bw: body weight; d: days; i.p.: intraperitoneal; MTD: maximum tolerated dose; ppm: parts per million; TCDD: 2,3,7,8-tetrachlorodibenzo-*p*-dioxin

[a] Data restricted to substances with exclusively negative results in transgenic mouse assays. ***

evaluation. Data on rats were not included due to the small number of studies.

Table 19. Characteristics of the Muta™Mouse assay and the Big Blue® mouse assay for predicting carcinogenic effects[a]

Term[b]	Calculation for Muta™Mouse	Calculation for Big Blue® mouse
Sensitivity	81% (25/31)	75% (18/24)
Specificity	100% (1/1)	100% (1/1)
Positive predictivity	100% (25/25)	100% (18/18)
Negative predictivity	14% (1/7)	20% (1/5)
Overall accuracy	81% (26/32)	76% (19/25)

[a] Weak positive results in transgenic mouse assay were judged as positive.
[b] Sensitivity = % of carcinogens with a positive result in the transgenic mutation assay (TMA); specificity = % of non-carcinogens with a negative result in TMA; positive predictivity = % of positive results in the TMA that are carcinogens; negative predictivity = % of negative results in the TMA that are non-carcinogens; overall accuracy = % of chemicals tested where TMA results agree with the carcinogenicity results.

Although the data pool is not comprehensive, some valuable information can be extracted from Table 19. Concerning the terms sensitivity and positive predictivity, both test systems (Muta™-Mouse and Big Blue®) are comparable and show a good predictivity for carcinogenicity.

As indicated by the positive predictivity in Table 19, there is 100% agreement between positive results in transgenic assays and in long-term carcinogenicity tests. Thus, a positive result in the transgenic assays strongly indicates that the substance tested is carcinogenic. The sensitivity is somewhat lower, because five non-genotoxic carcinogens that gave correct negative results in the transgenic animal mutagenicity assays (two for Muta™Mouse and three for Big Blue® mouse) and six false-negative results (four for Muta™Mouse and two for Big Blue® mouse) are included in this calculation.

Unfortunately, only two substances with negative results in long-term carcinogenicity studies are included in the comparison: one was tested in Muta™Mouse and one in Big Blue® mouse. Therefore, the terms specificity and negative predictivity cannot be adequately evaluated.

To summarize briefly, the transgenic mouse systems seem to be useful tools for detecting carcinogenic effects of a substance and for identifying a genotoxic mechanism of action.

10.2.4 Conclusion

The available data suggest that the sensitivity and positive predictivity of the transgenic assays for carcinogenicity are high. Most carcinogenic substances gave positive results in transgenic animal mutagenicity assays. The few non-carcinogens investigated in transgenic animal assays showed no mutagenic activity. Furthermore, there are substances for which transgenic assays indicate evidence for genotoxic mechanisms in carcinogenicity in contrast to other genotoxicity studies with conflicting or negative results.

For several substances, positive carcinogenicity results were obtained in rodents, but negative results were obtained in transgenic assays using the same species. There are several reasons for this disagreement:

1) The substance is more clastogenic than gene mutagenic (clastogenicity is not detected by Big Blue® mice or rats and Muta™Mouse).
2) The substance is a non-genotoxic substance giving a correct negative result in the transgenic animal mutagenicity assay.
3) The study design of the transgenic animal mutagenicity assay was suboptimal.

PART III:

APPLICATIONS OF TRANSGENIC ANIMAL MUTAGENICITY STUDIES

11. MUTATION FREQUENCIES AND SEQUENCING DATA AND APPLICATIONS OF THIS INFORMATION IN MECHANISTIC STUDIES

Although a primary purpose of transgenic animal mutagenicity assays is to examine whether a compound has genotoxic properties, the mutation assays can also be used to obtain important information regarding the nature of mutation. This chapter contains a brief overview of studies that have been carried out using transgenic mutation models and the information that has been derived from such studies.

An important feature of the transgenic animal mutagenicity assays is that mutations in the transgenic reporter genes are neutral and are not selected for or against in the rodent. This allows a researcher to study the genesis of mutations in the absence of their selection, expression and detection. In addition, molecular analysis of the mutations can reveal qualitative information regarding both spontaneous and induced mutation. Normally, this is achieved by isolating the mutant plaques, amplifying the DNA sequence of the target genes using polymerase chain reaction and determining the DNA sequence of relevant portions of the reporter gene (see section 6.2.10).

11.1 Studies on spontaneous mutant/mutation frequencies (in organs of non-exposed transgenic animals)

11.1.1 Sources of spontaneous mutations

In all organisms, spontaneous mutations arise from a variety of endogenous cellular processes. The following are recognized as primary sources of spontaneous mutations: 1) errors of DNA polymerases during replication, which result primarily in base pair substitution and short frameshift mutations; 2) deamination of cytosine and 5-methylcytosine, to form uracil and thymine, respectively; 3) oxidation of guanine to miscoding products such as 8-oxoguanine; 4) depurination, which results in miscoding abasic sites; 5) DNA strand breaks, which can lead to deletions and chromosomal rearrangements, and 6) mistakes in homologous recombination (i.e. during meiosis or V(D)J recombination), leading to deletions and chromosomal rearrangements (Friedberg et al., 1995).

11.1.2 Spontaneous mutation data: sequence data in organs of non-exposed transgenic animals

The intention of the transgenic mutation assays is to detect, quantify and characterize mutations arising in mouse or rat cells (i.e. in vivo mutations). There is, however, also the possibility that mutations can arise during the in vitro portion of the assay (i.e. in vitro mutations). Mutations that arise in vitro from DNA damage present in vivo are called ex vivo mutations. That such mutations can exist has been shown for the *lacI* and ΦX174 assays (Paashuis-Lew et al., 1997; Bielas & Heddle, 2000; Valentine et al., 2004). In the selective assays for *cII*, *lacZ* and *gpt* delta, however, the mutations are thought to be recessive, so that ex vivo mutations do not survive selection.

The *lacZ*, *lacI* and *cII* transgenes exhibit very similar high levels of spontaneous mutant frequencies, in the 10^{-5} range in most tissues (de Boer et al., 1998). Approximately 3500 independent spontaneous mutations have been examined using the Big Blue® transgenic mouse. Base substitutions predominate, although 16% of somatic and germline mutations are microdeletions, microinsertions or deletions combined with insertions. The *lacI* transgene shows similarity to the human *p53* gene in the pattern of microdeletions and microinsertions and the size distribution of microdeletions (Halangoda et al., 2001). In all tissues, the majority of spontaneous mutations are G:C→A:T transitions, which arise primarily at 5'-CpG-3' sequences, methylation sites that yield 5-methylcytosine. The transgenes are highly methylated in mammalian cells at 5'-CpG-3' sequences. Deamination of 5-methylcytosine yields thymine, which specifies the incorporation of adenine during DNA synthesis. The bulk of these studies have examined mutations in the *lacI* gene — and, more recently, the *cII* gene — of Big Blue® mice or rats; however, similar conclusions regarding the consistency of mutational spectra across somatic tissues and the importance of G:C→A:T transitions at 5'-CpG-3' sequences can be drawn from more limited sequencing of the *lacZ* gene in Muta™Mouse. Because of the high background of base substitutions, rare mutations such as small deletions are not readily quantified by the selections (Harbach et al., 1999).

Similar results were obtained in mutants from untreated *gpt* delta mice (Nohmi & Masumura, 2004). G:C→A:T transition mutations are the most prominent mutations; more than half of these occur at 5'-CpG-3' sites. G:C→T:A transversions are also frequently observed in the spontaneous *gpt* mutants. The remaining mutants contain frameshifts or short deletions.

In *gpt* delta rodents, the spontaneous Spi^- mutation spectrum is unique, in that most mutations are −1 deletions in repetitive sequences. There are several hotspots of spontaneously occurring Spi^- mutations. It has been suggested that these events are most likely induced by slippage errors of DNA polymerases during DNA replication. Although large deletions have been detected in both untreated and treated mice, specific hotspots for these events have not been characterized. This may indicate that double-strand breaks in DNA are randomly induced in the transgene region.

In the *lacZ* plasmid mouse, the spectrum of spontaneous point mutations has been determined in brain, heart, liver, spleen and small intestine. G:C→A:T transitions and 1 bp deletions were the predominant mutations, as has been observed in other systems. However, there was an observed difference in the mutation spectrum in young animals compared with older animals. This is discussed in more detail in section 11.1.3.

For the ΦX174 transgene *A*, sequencing can also confirm the in vivo origin of a mutation by demonstrating that most mutants from a large burst contain the same mutation (Valentine et al., 2004).

11.1.3 The frequency and nature of spontaneous mutations versus age in multiple tissues

Many tissues from different strains of transgenic mice with either a *lacI* or *lacZ* reporter gene have been assayed for spontaneous mutant frequencies. The factors that affect the inferred mutation rate are site of integration of the transgene, age, tissue and strain. About half of all mutations arise during development (and half of these in utero). In a study to assay the mutant frequencies from before birth to 28 days after birth, the F1 mice generated by crossing SWR females with Muta™Mouse males were assayed (Paashuis-Lew & Heddle, 1998). Analyses involved the evaluation of spontaneous mutant frequencies in entire embryos up to and including birth and were restricted to the small intestine for developmental stages following birth. The data showed that, as expected, many mutations arise early in development, by 12.5 days after conception. About one third of the mutations arise before birth, about one third occur during growth to adulthood and the remainder occur during the rest of the animal's life, depending on the tissue.

The steady-state level of spontaneous mutations in adult mice reflects the balance between the occurrence of new mutations and the elimination of mutated cells by selection (Nishino et al., 1996). In the majority of transgenic animal systems, mutations in the reporter genes are neutral, and there will be no selection of these mutations in any tissue. Several studies have examined the frequency and nature of spontaneous mutations versus age in multiple tissues (e.g. Nishino et al., 1996; Buettner et al., 1997; Hill et al., 2003, 2004). Based on these studies, the following conclusions have been made regarding spontaneous mutation.

Mutation frequencies showed tissue-specific increases with age but do not vary significantly from early to mid-adulthood. The time course of mutation frequency with age had significantly different shapes in different tissues. From 10 days to 3 months, mutation frequency increased significantly in liver and showed an increasing trend in cerebellum, forebrain and thymus. However, from early to mid-adulthood (3–10 months), there was no significant further increase in mutation frequency in any of the tissues evaluated: brain (whole brain, cerebellum and forebrain), thymus, liver, adipose tissue and male germline. From 10 to 25 months, the mutation frequency increased significantly in liver and adipose tissue, but not in cerebellum, forebrain and the male germline.

Mutation frequencies were generally low in the male germline. The mutation frequency in the male germline was consistently the lowest, remaining essentially unchanged in old age.

The spectrum of mutation types was similar with age and tissue type. It did not vary with differences in gender or mouse genetic background. A minor class of mutations, tandem-base substitutions, is unique in having marked tissue, age and spectral specificity (Buettner et al., 1999; Hill et al., 2003). The mechanism and significance of this observation are unclear.

Somewhat different results have been obtained using the *lacZ* plasmid assay. Dollé et al. (1997, 2000) reported that both the frequency and molecular nature of spontaneous mutations were dependent on the tissue and the age of the animals. In the liver and heart of young animals, half of the spontaneous mutations were size change mutations, while in the small intestine, only one third were size change mutations. In old animals, however, 3–4 times more point mutations than size mutations were observed in brain and small intestine, whereas in the liver, the distribution between size changes and point mutations remained equal.

In the *lacZ* plasmid mouse, the spectra of spontaneous point mutations were determined in brain, heart, liver, spleen and small intestine. G:C→A:T transitions and 1 bp deletions were the predominant mutations. This similar mutant spectrum observed at a young age may reflect a common mutation mechanism for all tissues that could be driven by the rapid cell division that takes place during development. In old animals, a strong increase in G:C→A:T transitions was observed in the slowly dividing tissues brain, heart and liver. In small intestine, G:C→A:T transitions were also observed, and the frequency of G:C→T:A, G:C→C:G, all base substitutions involving A:T base pairs and 1 bp deletions was increased. Apparently, the spectra of the young tissues did not resemble that of a highly proliferative aged tissue, small intestine, implying that differences in organ function, possibly associated with the proliferative capacity of the tissue, may explain the divergence in mutation spectra during ageing (Dollé et al., 2002).

11.2 Examination of fundamental paradigms in genetic toxicology

Several studies have used transgenic animals to examine fundamental paradigms in genetic toxicology. Here, we summarize some studies with transgenic animals that have addressed the issues of 1) dose–response relationship of genotoxic carcinogens and 2) the relationships among DNA adduct formation, mutation frequency and cancer in rodents.

11.2.1 Dose–response relationships

Humans are chronically exposed to most environmental chemicals at low doses. However, genotoxicity assays are usually performed at high doses with short treatment periods. A dose–response relationship in transgenic mutation assays has been reported after single and repeated exposure in both Big Blue® and Muta™Mouse assays. Different dose levels at the same exposure period have been used, as well as different exposure periods with the same single or repeated daily dose.

Topinka et al. (2004a) administered a single dose of cyproterone acetate by gavage to female Big Blue® rats at dose levels of 0, 5, 10, 20, 40, 80 or 160 mg/kg of body weight. The authors demonstrated a dose–effect relationship for mutation frequency in *lacI* in the liver (Fig. 11). The statistical analysis revealed a significant effect at a dose greater than or equal to 10 mg/kg of body weight. A dose of 5 mg/kg of body weight was ineffective. However, this dose resulted in a 2.5-fold increase in the mutation frequency when multiple dose treatment (0, 0.1, 1, 5 mg/kg of body weight per day for 3 weeks) was used with a similar experimental design (Topinka et al., 2004b).

A dose–response that is non-linear has been suggested by Tombolan et al. (1999a). 5,9-Dimethyldibenzo[*c,g*]carbazole was administered once by topical injection at doses ranging from 3 to 180 mg/kg of body weight, and the liver was sampled 28 days later. The mean mutant frequency was slightly increased at 3 and 10 mg/kg of body weight and markedly increased at 30, 90 and 180 mg/kg of body weight. The possible influence of cell proliferation on this effect is discussed below (section 11.2.3).

The currently accepted view concerning mechanisms of carcinogenicity is that no threshold exists for carcinogenic effects of genotoxic carcinogens. In a feeding study, Hoshi et al. (2004) tested low doses (0.001, 0.01, 0.1, 1, 10 or 100 mg/kg in the diet for 16 weeks) of the heterocyclic amine 2-amino-3,8-dimethylimidazo[4,5-*f*]quinoxaline (see also Master Table, Appendix 1) in Big Blue® rats. The frequency of *lacI* mutants as well as glutathione *S*-transferase positive foci (indicating preneoplastic hepatocytes) in the liver were determined. Positive foci significantly increased at 100 mg/kg. Significantly increased mutation frequencies were reported at 10 and 100 mg/kg; no effects were observed at 0.001–1 mg/kg. The DNA sequence analysis revealed a very characteristic mutation spectrum produced by 2-amino-3,8-dimethylimidazo[4,5-*f*]quinoxaline at higher doses. The authors suggested that their data demonstrated a no-observed-effect level (NOEL) for both preneoplastic lesions and mutagenicity; in addition, it was suggested that the NOEL for mutagenicity was lower than that for preneoplastic lesions.

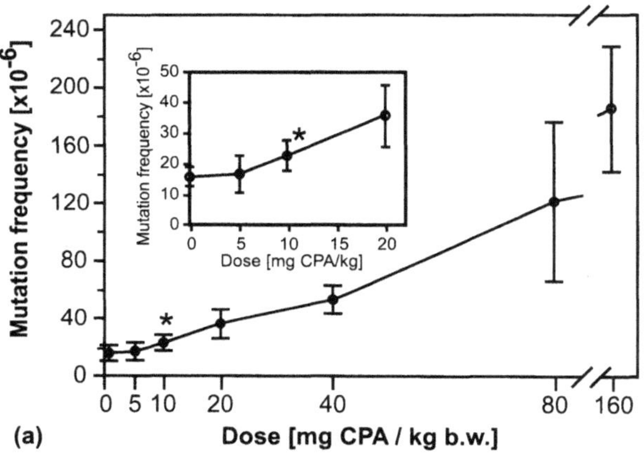

Fig. 11. Dose dependence of mutation frequencies in *lacI*. Groups of five transgenic rats were treated with single doses of cyproterone acetate (CPA) as indicated, and mutation frequencies were determined 2 weeks later. Mean values ± standard deviation are shown. b.w. = body weight. *: significance level P = 0.027. (Reprinted from Topinka et al., 2004a, with permission from Elsevier)

A study of dose–response relationships with a weak mutagen was carried out by de Boer et al. (1996). Male B6C3F1 Big Blue® mice were treated by gavage with tris(2,3-dibromopropyl)phosphate at dose levels of 0, 150, 300 or 600 mg/kg of body weight per day for 2 (low dose) or 4 days (middle and high dose). A slight increase of 50% over the control value for mutant frequency of *lacI* transgene was detected in the kidney (also the main target organ for carcinogenicity in male mice) after the high-dose exposure, but no effects were observed in stomach and liver. Statistical analysis revealed significance of this effect in the kidney at the level $P < 0.01$. The DNA sequence analysis suggested a treatment-related dose-dependent change in the mutation spectrum in the kidney that was characterized primarily by the loss of single G:C base pairs, even at the low and middle dose levels. These results illustrate the sensitivity and the specificity of the Big Blue® mouse assay in the analysis of target organ mutation.

Overall, a dose–response relationship can be shown with transgenic mutation assays. Furthermore, it has been demonstrated that the sensitivity of the transgenic assay can be increased by measuring not only the parameter mutation frequency, but also sequence analysis.

11.2.2 Correlation of dose with mutation frequency and carcinogenicity

A correlation between the mutation frequency of *lacI* (Suzuki et al., 1996b) and cancer incidence (Nagao et al., 1998, 2001) of the heterocyclic amine 2-amino-3,4-dimethylimidazo[4,5-*f*]quinoline has been reported in female Big Blue® mice of the strain C57BL/6. Exposure to 2-amino-3,4-dimethylimidazo[4,5-*f*]quinoline in the diet at 300 mg/kg for a period of 92 weeks induced an increased tumour incidence predominantly in liver and large intestine (caecum and colon). After 12 weeks of exposure to the same dose, an increased mutation frequency was detected in *lacI* of colon and liver, the colon showing an increase approximately 8-fold higher than the liver, which might be related to the higher proliferation rate in the colon. The colon, but not the liver, showed increased mutation frequencies even after 1 week of exposure to 2-amino-3,4-dimethylimidazo[4,5-*f*]quinoline.

In several transgenic studies, the dose level used is comparable to the dose levels used in long-term carcinogenicity studies on the same species. For example, in inhalation studies, 1,2-dibromoethane significantly increased tumour incidence in lung and nasal cavity of mice and rats at dose levels of 80 and 310 mg/m^3 (IARC, 1999d). In an inhalation study using male Muta™Mouse (Schmezer et al., 1998a), a single exposure to 230 mg/m^3 for 2 h did not increase the mutant frequency in *lacZ* of these two target organs after a 14-day post-exposure observation period. In contrast, repeated daily 2-h inhalation treatment for 10 days with the same concentration (sampling time 14 days after the last exposure) resulted in a significant increase in the mutant frequency in nasal mucosa (no effect in the lung).

The antiestrogenic drug tamoxifen has been used in the therapy of breast cancer. In long-term gavage studies on rats (IARC, 1996), this substance induced liver tumours in males and females even at the lowest dose tested, when the experimental design involved treatment with 0, 5, 20 or 35 mg/kg of body weight per day for 2 years. In female Big Blue® rats, a dose-dependent increase in the mutation frequency of *lacI* was detected in the liver after daily gavage with 0, 10 or 20 mg/kg of body weight for 6 weeks (Davies et al., 1997).

11.2.3 Relationship between DNA adducts, cell proliferation and gene mutations

Comparisons of different end-points are complicated by the kinetics that these end-points display. For example, the DNA adducts responsible for mutations in a tissue may arise and be detectable very soon after treatment and subsequently decline as adducts are repaired or converted to mutations. In contrast, the mutations that arise from them increase only relatively slowly as the tissue turns over. Comparisons between the frequencies of these two end-points at any one time will thus be misleading. In order to relate adducts to mutations, the two end-points should be measured at the separate times at which they are at their maxima. Similar problems exist for many of the interesting comparisons between kinetically distinct end-points and responses in vivo, including carcinogenicity.

Cell turnover is considered a critical factor in the conversion of DNA adducts into mutations. Tombolan et al. (1999a) used the potent mouse liver carcinogen 5,9-dimethyldibenzo[*c,g*]carbazole to examine the kinetics of induction of DNA adducts, cell proliferation and *lacZ* gene mutations in the liver of the Muta™Mouse after single topical application of 10 or 90 mg/kg of body weight and post-exposure observation periods of 2, 4, 7, 14, 21 and 28 days. Both doses induced a similar level of persistent DNA damage. However, the mutant frequency was increased only 2-fold after the low dose, whereas the high dose induced a marked 44-fold increase. These differences between DNA adducts and mutant frequency might be related to the kinetics of the proliferation rate: no change in proliferation rate was detected at the low dose, but at 90 mg/kg of body weight, a regenerative cell proliferation was observed. These results suggested that regenerative cell proliferation induced by the high dose allowed the 5,9-dimethyldibenzo[*c,g*]carbazole-induced DNA adducts to be fixed as stable mutations. In a subsequent study (Tombolan et al., 1999b), it was shown that the low dose followed by induction of regenerative cell proliferation by carbon tetrachloride also resulted in a marked increase (15-fold control versus 2-fold increase with 5,9-dimethyldibenzo[*c,g*]carbazole alone at 10 mg/kg of body weight). Similar results were demonstrated with an additional proliferative treatment using the mitogenic agent phenobarbital.

The relationship between DNA adducts, mutation and cell proliferation has also been studied using the antiandrogenic drug cyproterone acetate in the liver of Big Blue® rats (Wolff et al., 2001; Topinka et al., 2004a). A dose-dependent induction of DNA adducts and *lacI* mutations 6 weeks after a single oral dose at dose levels of 0, 25, 50, 75, 100 and 200 mg/kg of body weight was reported (Fig. 12).

In these experiments, the highest non-effective dose for mutagenicity is 50 mg/kg of body weight, although 25 and 50 mg/kg of body weight induced high levels of DNA adducts, again suggesting that proliferation is strongly involved. The authors have demonstrated that the mitogenic activity of cyproterone acetate itself triggers the expression of cyproterone acetate–specific mutations. The emergence of S-phase cells and of mitotic figures in the liver of transgenic rats 1, 2 or 3 days after an oral cyproterone acetate dose of 40 or 160 mg/kg of body weight was studied. Twenty-four hours after application of the high dose, a maximum of approximately 50% S-phase cells was reached (with the low dose, this was approximately 20%); 24 h later, mitoses attained a maximum (a 6-fold increase). The DNA synthesis rate reached the control level at the third day. It is concluded that the low endogenous proliferation rate of the liver did not contribute significantly to the expression of mutations, but doses above 50 mg/kg of body weight are necessary for induction of a mitogenic activity sufficient for conversion of DNA adducts into mutations detectable 6 weeks after application.

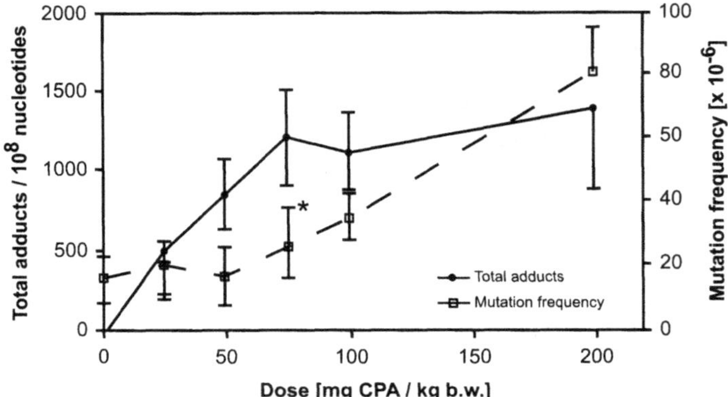

Fig. 12. Dose dependence of DNA adduct levels and mutation frequencies, 6 weeks post-exposure. Groups of five female Big Blue® rats were treated with single oral doses of cyproterone acetate (CPA) as indicated in the figure. Mean values ± 95% confidence levels are shown. b.w. = body weight. *: Significantly differing from controls ($P < 0.03$). (Adapted from Figure 2 in Wolff et al. (2001). Reprinted with kind permission of Springer Science and Business Media.)

In further experiments, the authors studied the time course of the DNA adduct levels and of *lacI* mutation frequencies in the liver 1, 2 and 3 days and 1, 2, 4, 6 and 8 weeks after a single oral cyproterone acetate dose of 100 mg/kg of body weight. The DNA adduct levels strongly increased in the first 3 days and subsequently reached a steady state 8 weeks after application at approximately 40% of the maximum. The time course of the mutation frequency was characterized by a strong increase within 2 days after application followed by a plateau that remained up to day 14 post-exposure. Thereafter, the mutation frequency decreased by about 80% within 2 weeks (DNA adducts decreased only 13% in this period) and was maintained at this low level for a further 4 weeks. The authors suggested that this rapid reduction in mutation frequency is due to a specific elimination of liver cells carrying mutation (including *lacI* mutations), but this would mean that the mutations are not genetically neutral. This reduction in mutation frequency seems to be complete within 4 weeks after dosing, which would be consistent with the turnover of liver cells as seen from the manifestation time for the liver (Douglas et al., 1996). The decrease in mutation frequency in the liver at later sampling times observed in this study was not observed after exposure to *N*-ethyl-*N*-nitrosourea (Douglas et al., 1996) or some other chemicals.

The relationship between DNA adduct levels and mutant frequency in *lacI* was examined in various organs of female Big Blue® mice after exposure to 2-amino-3,4-dimethylimidazo[4,5-*f*]quinoline at 300 mg/kg via the diet for 12 weeks (Ochiai et al., 1998; Nagao et al., 2001) using the same tissue samples. The highest DNA adduct levels were detected in the liver (28.3 mol/10^7 molecules), followed by heart (8.4), colon (3.3), forestomach (1.3) and bone marrow (0.4). The mutant frequency was more elevated in the colon (38-fold control), followed by bone marrow (5.7-fold), liver (4.6-fold) and forestomach (2.5-fold) (heart: no effect). Thus, there is no direct correlation between adduct levels and mutant frequency.

11.3 Studies into the mechanism of action of mutagenicity/carcinogenicity using sequence data

The ability to sequence induced mutations in transgenic reporter genes provides an investigator with important information regarding several important aspects of mutation. The following studies are examples that demonstrate how transgenic animal mutagenicity assays and subsequent spectral analysis can be used to examine different aspects of the activity of mutagenic agents.

11.3.1 *Clonal correction and correction for ex vivo mutations*

DNA sequencing can be used to obtain more accurate estimates of both spontaneous and induced mutations in those cases where high interanimal variation is observed. In particular, it is useful in these cases to exclude the possibility that clonal events or "jackpots" are responsible for the observed variation and to correct the mutant frequency for the particular animal. This requires the removal from the data set of all but one mutant derived from a specific type of mutation at a single site for each tissue of an animal and subsequent correction of the mutant frequency for those mutants that are discarded.

Although it is extremely rare for ex vivo mutations to be observed using the positive selection systems used in most transgenic animal models at the current time, ex vivo mutations are more common in the ΦX174 system. Such mutations can be identified using sequencing: multiple plaques from a single, large burst contain the same mutation (Valentine et al., 2004).

11.3.2 Premutagenic lesions

Induced mutations generally arise from erroneous replication or repair of DNA lesions. Many studies have demonstrated that the nature of a DNA adduct strongly influences mutagenic outcome. The following are examples of studies that have linked DNA lesions induced by different chemicals to different mutational outcomes.

In *lacZ* transgenic mice, treatment with *N*-ethyl-*N*-nitrosourea, followed by sampling of mutations in liver and bone marrow, showed that A:T→T:A transversions and A:T→G:C transitions were prominent in both liver and bone marrow of *N*-ethyl-*N*-nitrosourea-treated mice, suggesting the involvement of unrepaired O^2- and O^4-ethylthymine adducts (Douglas et al., 1996). This contrasts with the G:C→A:T transitions in 5'-CpG-3' sites characteristic of untreated mice. In further studies of different alkylating agents (*N*-ethyl-*N*-nitrosourea, diethylnitrosamine and ethyl methanesulfonate), comparative analysis of data on adducts (O^6-ethylguanine and N^7-ethylguanine), mutation induction and mutation spectra from the *lacZ* transgene provided information regarding the mutagenicity of individual adducts in different tissues (Mientjes et al., 1998).

Lynch et al. (1998) investigated the mutagenicity of 2-amino-1-methyl-6-phenylimidazo[4,5-*b*]pyridine in the Muta™Mouse. Thirty-three per cent of the mutants from the treated group showed G:C→T:A transversions from a total of 65% base substitutions compared with 17% in the vehicle control group. Twenty per cent of the 2-amino-1-methyl-6-phenylimidazo[4,5-*b*]pyridine group mutants were due to G:C base pair (−G) deletions (none in control). The observed mutational spectrum was consistent with the known effects of the principal dG–8 adduct induced by 2-amino-1-methyl-6-phenylimidazo[4,5-*b*]pyridine, as determined using a variety of other mutational systems in vivo and in vitro.

Using the Muta™Mouse, it was shown that *o*-aminoazotoluene dG–8 adduct increased the mutant frequency of the integrated target gene (*lacZ*) in liver (3.4-fold) and colon (6.5-fold) (Ohsawa et al., 2000). The mutant frequency in the *λcII* in liver and colon was found to be 5 and 9 times higher, respectively, in *o*-aminoazotoluene-treated mice than in control mice. Sequence analysis in *cII* revealed that *o*-aminoazotoluene induced G:C→T:A transversions, whereas spontaneous mutations consisted primarily of G:C→A:T transitions at CpG sites (Kohara et al., 2001). dG adducts of *o*-aminoazotoluene were generated in the mouse *cII* gene and resulted in G:C→T:A transversions.

7,12-Dimethylbenz[*a*]anthracene is metabolically activated to form DNA adducts with dG and dA (RamaKrishna et al., 1992). These adducts or depurination products cause misincorporations during DNA synthesis, which lead to G:C→T:A transversions or A:T→T:A transversions. The latter are rare in spontaneous mutations. This compound has been investigated using molecular analysis of in vivo *lacI* mutations in Big Blue® rats (Mittelstaedt et al., 1998; Manjanatha et al., 2000; Shelton et al., 2000). DNA sequencing revealed that the majority of 7,12-dimethylbenz[*a*]anthracene-induced *lacI* mutations were base pair substitutions and that A:T→T:A (48%) and G:C→T:A (24%) transversions were the predominant types.

Investigation into the spectra of mutations induced by the carcinogenic pyrrolizidine alkaloid riddelline in the liver *cII* gene of transgenic Big Blue® rats showed a statistically significant difference between the spectra of mutations of treated and control rats (Mei et al., 2004). A G:C→T:A transversion (35%) was the major type of mutation in rats treated with riddelline, whereas a G:C→A:T transition (55%) was the predominant mutation in the controls. Treated rats showed an unusually high frequency (8%) of tandem base substitutions of G:G→T:T and G:G→A:T. These results indicate that riddelline is a genotoxic carcinogen in rat liver and that types of mutations induced by this compound are consistent with riddelline adducts involving G:C base pairs.

Molecular analysis of in vivo *cII* gene mutations in the mammary tissue of female transgenic (Big Blue® F344 × Sprague-Dawley)F1 rats treated with 6-nitrochrysene has shown that the structures of 6-nitrochrysene–DNA adducts are consistent with the mutational spectra (Boyiri et al., 2004). Control mutants consisted primarily of G:C→A:T transitions, whereas 6-nitrochrysene-induced mutants were composed of several major classes of mutations, with G:C→T:A, G:C→C:G, A:T→G:C and A:T→T:A as the most prevalent. Both incidence and multiplicity of mammary adenocarcinomas were significantly elevated at the highest dose.

11.3.3 Tissue-specific responses

As mentioned in section 10.1, not all organs with a high rate of induction of mutation in the *lacZ* transgene develop tumours in Muta™Mouse (Hakura et al., 1998). In further studies (Hakura et al., 1999, 2000), the mutational spectra of the *lacZ* transgene were compared in two target organs for carcinogenicity (forestomach and spleen) and two non-target organs (colon and glandular stomach) obtained 2 weeks after five daily consecutive oral treatments with benzo[*a*]pyrene at 125 mg/kg of body weight per day. All these organs were highly mutated in the *lacZ* transgene. The sequence data showed similar mutational spectra of the *lacZ* transgene in the two target organs; the predominant mutations were G:C→T:A transversions (55% and 50% for forestomach and spleen, respectively), followed by deletions (20% and 21% for forestomach and spleen, respectively), mainly at the G:C site. In contrast, the mutational spectra of the *lacZ* transgene in the two non-target organs were significantly different from those in the target organs (G:C→A:T transitions were found) and also differed from one another in the incidence of G:C→T:A transversions and deletions (Hakura et al., 2000).

Aristolochic acid is part of a mixture of nitrophenanthrene derivatives found in several plant species and used as a herbal drug (Kohara et al., 2002a). To evaluate the in vivo mutagenicity of this compound, the mutant frequency was analysed in the *lacZ* and *cII* genes of 10 organs of Muta™Mouse after gavage. The nature of the mutations induced by aristolochic acid was investigated by sequence analysis of the *cII* gene. The mutant frequencies in the target organs (forestomach, kidney and bladder) of the treated mice were significantly higher than those of the control mice, whereas the mutant frequencies in non-target organs, except the colon, showed only slight increases. Sequence analysis of *cII* mutants in target organs revealed that aristolochic acid induced mainly A:T→T:A transversions, whereas G:C→A:T transitions at CpG sites predominated among spontaneous mutations. The results suggest that aristolochic acid, which is activated by cytochrome P450 and peroxidase to form cyclic nitrenium ions, causes the A:T→T:A transversions in the target organs of the mice by forming dA adducts.

The mutational spectrum of 2-amino-3,4-dimethylimidazo[4,5-*f*]quinoline was investigated in the liver and bone marrow of transgenic mice carrying the *lacI* gene. In the liver, G:C→T:A transversions were the most frequent events, accounting for 46% of the total mutations, followed by G:C→A:T transitions (25%). In the bone marrow, four types of mutations, G:C→T:A transversions, G:C→A:T transitions, complex mutations and single base deletions, each accounted for 21–23% of the total mutations. Control mice showed frequent G:C→A:T transitions at CpG sites. The results suggest a tissue-specific mechanism of mutagenesis (Ushijima et al., 1994).

N-Hydroxy-2-acetylaminofluorene administered to Big Blue® rats in multiple doses produces *N*-(deoxyguanosin-8-yl)-aminofluorene as a major adduct in the target liver and in two non-target tissues (bone marrow cells and spleen lymphocytes) (Chen et al., 2001a). Increased mutant frequencies were noted in both the *Hprt* and *lacI* genes of spleen lymphocytes, and there were about 10-fold more *lacI* mutations in the liver than in splenic lymphocytes. The *lacI* mutant frequencies in the tissues of treated rats correlated with the extent of DNA adduct formation. Sequence analysis was conducted on *lacI* DNA and *Hprt* cDNA from the mutants, to determine the mutational specificity of *N*-hydroxy-2-acetylaminofluorene in the rat. All the mutation spectra differed significantly from the corresponding mutation profiles from the untreated animals. G:C→T:A transversion was the most common mutation in all mutation sets. However, there were significant differences in the patterns of base pair substitution and frameshift mutation between liver and spleen *lacI* mutants and between spleen lymphocyte *lacI* and *Hprt* mutants, suggesting differences in the nature of mutation in target and non-target organs.

11.3.4 Evaluation of genotoxicants that do not appear to interact with DNA

In chronic exposures, compounds may be genotoxic despite the fact that they do not induce any direct damage in DNA. In the case of at least two chemicals, phenobarbital (Shane et al., 2000c) and oxazepam (Singh et al., 2001), DNA sequence analysis has revealed a spectrum of mutations that is consistent with that of oxidative damage in DNA that is a result of induction of cytochrome P450 2B isozymes by chronic administration of cytochrome P450 2B inducers (Shane et al., 2000c).

11.3.5 Active components of mixtures

Different compounds or classes of compounds often produce distinctive mutational spectra; thus, it may be possible to use sequence analysis to determine the mutagenic components present in a mixture. For instance, coal tar is a complex mixture of aromatic and aliphatic hydrocarbons. In the *lacZ* gene of Muta™Mouse, the mutational spectrum of coal tar has been found to induce primarily G:C→T:A transversions and 1 bp deletions of G:C base pairs (Vogel et al., 2001) — a spectrum that was very similar to that of benzo[*a*]pyrene (Hakura et al., 2000) in the same gene. This implicates benzo[*a*]pyrene and related polycyclic aromatic hydrocarbons as the active components of mutagenic coal tar.

11.3.6 Active metabolites

Mutation induction by tamoxifen and α-hydroxytamoxifen in the liver *cII* gene of Big Blue® rats has been assessed, and the types of their mutations induced in the liver *lacI* and *cII* genes have been characterized. Molecular analysis of the mutants showed that the α-hydroxytamoxifen-induced mutational spectrum differed significantly from the control spectrum, but was very similar to the spectrum induced by tamoxifen in both *lacI* and *cII* genes. G:C→T:A transversion was the major type of mutation in both the treated samples, while G:C→A:T transition was the main type of mutation in the control. These results support the hypothesis that α-hydroxytamoxifen is a major proximate tamoxifen metabolite causing the initiation of tumours in the liver of rats treated with tamoxifen (Chen et al., 2002).

11.3.7 Investigations into the mechanisms of deletion mutations in vivo

Using Spi$^-$ selection, deletion mutations can be analysed in the λ phage of the *gpt* delta mouse carrying about 160 copies of λEG10 DNA per diploid. In the following studies, sequence characteristics of Spi$^-$ mutants recovered from *gpt* delta mice were analysed and the mechanisms of deletion mutations investigated. Such studies have included mutation induction by ionizing radiation in liver and spleen (Nohmi et al., 1999; Masumura et al., 2002; Yatagai et al., 2002), ultraviolet B (UVB) in epidermis (Horiguchi et al., 2001), mitomycin C in bone marrow (Takeiri et al., 2003) and the heterocyclic amine, 2-amino-1-methyl-6-phenylimidazo[4,5-*b*]pyridine, in the colon of *gpt* delta mice (Masumura et al., 1999b, 2000). It was found that different treatments induce different types of deletions. For example, ionizing radiation, UVB and mitomycin C treatment induce large deletions with sizes of more than 1–2 kb, whereas 2-amino-1-methyl-6-phenylimidazo[4,5-*b*]pyridine induces 1 bp deletions in G runs (Nohmi & Masumura, 2004, 2005). About half of the large deletions have short homologous sequences of 1–12 bp at the junctions of mutants. About 10% of the large deletions have insertions in the junctions. The length of the insertion sequence is usually 1 or 2 bp, but the maximum insertion is 14 bp. The results are consistent with the notion that ionizing radiation induces double-strand breaks through direct deposition of energy on DNA and through oxidation of DNA (Masumura et al., 2002; Yatagai et al., 2002), while UVB and mitomycin C induce lesions that block the progression of the replication fork, thereby inducing double-strand breaks in DNA (Horiguchi et al., 2001; Takeiri et al., 2003). UVB and mitomycin C induce pyrimidine dimers and interstrand cross-links in cDNA strands, respectively, which strongly block DNA replication (De Silva et al., 2000; Limoli et al., 2002). The resulting double-strand break ends can be digested by exonucleases, generating 3'- or 5'-protruding ends, followed by annealing of the complementary short homologous sequences, gap filling and ligation.

A major advantage of the *lacZ* plasmid transgenic reporter mouse model is its sensitivity for large deletions and genome rearrangements. It was demonstrated by Dollé et al. (1997, 2000) that next to point mutations and small deletions, larger rearrangements with one breakpoint in the *lacZ* plasmid and one elsewhere in the mouse genome also contribute to the mutant spectrum. As is the case with point mutations (Dollé et al., 2002), genome rearrangements are tissue- and age-dependent as well. High numbers of rearrangements were found in heart, liver and small intestine, and lower numbers in the brain (Dollé & Vijg, 2002).

Individual plasmids were obtained by digesting genomic DNA with the restriction enzyme HindIII, cutting once in each wild-type plasmid. The recovered mutant plasmids were characterized by restriction enzyme analysis on agarose gels. Mutant plasmids with restriction patterns identical to wild-type plasmids were categorized as no-change mutants and comprised gene mutations and small deletions. Mutants with altered restriction patterns were size change mutations. It was reported that 83% of the size change mutations were rearrangement with one breakpoint in the *lacZ* plasmid and one in the murine genome (Dollé & Vijg, 2002).

With the knowledge on sequence and exact chromosomal location of the *lacZ* plasmids and the mouse genome sequence database (Marshall, 2001), it became possible to obtain sequence information on the breakpoint regions. The sequence of 38 recovered mutants obtained from untreated mice was analysed (Dollé & Vijg, 2002). About 50% of these demonstrated breakpoint sequences in the *lacZ* reporter gene and one specific for the reporter gene chromosome at a distance from 100 kb up to 66 Mb. These mutants presumably represent intrachromosomal rearrangements. Taking into account the fact that exclusively genome rearrangements can be detected with an intact 5' sequence (upstream), since the origin of replication and the ampicillin resistance gene must be present, sequence studies demonstrated that half of the intrachromosomal rearrangement would have been inversions, whereas deletions and transpositions each made up one quarter. The other 50% of the mutants had one breakpoint sequence from the reporter gene chromosome and one from a sequence randomly from another chromosome. These were classified as translocations.

Chromosomal inversions are detected in the pKZ1 recombination assay. The inversions are mediated by mouse immunoglobulin recombination signals and are likely to occur via a non-homologous end joining process. Sequence analysis across the inversion breakpoint can be performed using polymerase chain reaction. In spleen and prostate, the breakpoints identified resemble those expected for

RAG1/RAG2-mediated recombination (Sykes et al., 1998; Hooker et al., 2004b). Non-linear dose–responses have been identified after low-dose exposure to a number of agents. This assay may be useful for studying the role of non-homologous end joining repair in complex dose–responses to mutagenic agents.

11.4 Importance of the transgenic mutation assays for studies other than genetic toxicology

A very important application of these transgenic mouse assays has been in fundamental studies on the origin of mutations and the roles of various biological processes in preventing them. These studies have included studies of the DNA repair mechanisms, carcinogenesis, ageing and inherited genetic conditions affecting these processes. This is a very active field of research, and these assays are playing an important part, as any current review of any of these topics will show. Transgenic mice (*lacI*, *lacZ*, *gpt* delta, plasmid *lacZ*) lacking DNA repair genes such as *p53*, *atm*, *parp-1* or *ogg1* have been successfully established and play important roles in elucidating the mechanisms suppressing the genome instability induced by endogenous and exogenous environmental stresses (Buettner et al., 1997; Klungland et al., 1999; van Oostrom et al., 1999; Minowa et al., 2000; Arai et al., 2002, 2003; Yatagai et al., 2002; Furuno-Fukushi et al., 2003; Wijnhoven & van Steeg, 2003; Hoogervorst et al., 2004; Shibata et al., 2005). Transgenic mice, for example pKZ1 mice, can be bred to tumour model mice to study the effect of cancer-associated genes on chromosomal changes in the early stages of carcinogenesis (Hooker et al., 2004b).

PART IV:

EVALUATION, CONCLUSIONS AND RECOMMENDATIONS

12. EVALUATION OF THE TRANSGENIC ANIMAL MUTAGENICITY ASSAYS BASED ON THIS REVIEW OF THE CURRENT LITERATURE

12.1 Features of the assay

In comparison with other in vivo genotoxicity tests, an important advantage of the transgenic animal mutagenicity assays is that there is theoretically no target tissue restriction. Mutation can be evaluated in all tissues for which sufficient amounts of DNA can be collected. Transgenic test systems are valuable tools for assessment of organotrophic effects in mutagenicity (Schmezer et al., 1998b) and can be used for analysis of mutagenic potency in organs with very different proliferative capacity (e.g. Hakura et al., 1998). In contrast, several existing in vivo assays are extremely limited with respect to the tissue: the mouse micronucleus is normally performed in the bone marrow or peripheral blood, while the UDS assay is normally limited to liver.

In the transgenic animal mutagenicity assay, test compounds can be administered through any route. This allows the researcher to use the most relevant mode of administration for a compound and evaluate tissues that are of concern. In addition, there are no restrictions subsequent to absorption and distribution of the test substance, as is the case, for example, in the mouse spot test, in which the chemical must pass through the placental barrier and the number of melanoblast target cells is extremely limited.

Local as well as systemic mutagenic effects may be detected with the transgenic animal system after dermal (skin painting), inhalation or oral exposure, and also after parenteral injection (Nohmi et al., 2000). There is no other in vivo test system available that detects gene mutations at the site of contact.

12.2 Gene mutation assay — implications for testing

To date, the transgenic animal mutagenicity assay has not been heavily used by industry in toxicological screening, in large part because an OECD Test Guideline has not yet been developed. Recently, an internationally harmonized protocol has been recommended (Thybaud et al., 2003), and this protocol should form the basis for such a guideline. The utility of such a guideline is based, in part, on the fact that the transgenic animal mutagenicity assays are capable of detecting gene mutations. If such a protocol is used, a negative result can be considered as reliable. As such, it is superior to the mouse spot test, which is rarely carried out anymore, and the UDS test, which measures DNA damage and is recognized as being a surrogate test for mutation.

The analyses carried out in this document demonstrate that the transgenic animal mutagenicity assays have a high positive predictivity for chemically induced tumorigenesis that is superior to that of the mouse bone marrow micronucleus assay. Importantly, the transgenic animal mutagenicity assays and the mouse micronucleus assay exhibited a high degree of complementarity. This complementarity may be attributed to the fact that together they detect both gene mutations and chromosomal aberrations and are therefore capable of detecting genotoxic chemicals that act through a wide range of genotoxic mechanisms.

A further benefit of transgenic animal mutagenicity assays is that they permit the confirmation of the results of in vitro gene mutation assays using the same end-point in vivo.

12.2.1 Reliability of a negative result

Provided that a suitable protocol was used, that tissue exposure can be demonstrated and that the appropriate tissues were sampled, a result that does not demonstrate a significant increase can be regarded as negative with confidence. Under certain circumstances, when scientifically justified, evaluation of a single tissue might be sufficient to define a negative. For a chemical for which the tissues at risk are not known from pharmacokinetic or other toxicological data, measurement of bone marrow, liver and a tissue relevant to the route of administration should be regarded as sufficient to establish a negative. For example, for an inhalation study, lung would be a suitable third tissue, whereas for oral administration, the gastrointestinal tract epithelial cells would be suitable.

12.3 Comparison with endogenous genes

In comparison with the few endogenous loci (e.g. *Dlb-1*, *Hprt*) at which mutation has been measured in vivo, the response of the transgenic reporter genes appears to be comparable for acute dosing, but not for chronic dosing. The spontaneous mutant frequency in transgenic animals is relatively high (in the mid 10^{-5} range in most tissues; Gorelick, 1995) compared with that in many endogenous loci, and this may reduce the sensitivity of the assay under acute treatments that are not currently regarded as appropriate to a well conducted transgenic animal mutagenicity assay. However, mutations at the transgene accumulate linearly with time, whereas mutations at the endogenous gene are much less frequent than those at the transgene. This is primarily a function of the neutral quality of the transgenes; in contrast, there is strong selective pressure acting against mutations in endogenous genes. Thus, under the conditions that are currently recommended for testing, the transgenic animal mutagenicity assays may provide a more sensitive estimate of the genotoxicity of a compound than is the case with endogenous loci that are currently available for evaluation.

It may be noted that in transgenic reporter genes, there is no transcription of transgenes and no transcription-coupled DNA repair, which might result in higher spontaneous and treatment-related mutation rates than in endogenous genes after chronic exposure to mutagens.

12.4 Molecular analysis and mechanistic studies

The transgenic animal mutagenicity assays are advantageous over many other in vivo genotoxicity tests, in that molecular analysis of mutations may be carried out relatively easily using DNA sequence analysis or Southern hybridization analysis. Such analyses have demonstrated that the transgenic animal mutagenicity assay is useful for detection of point mutations, small insertions and deletions, but is not generally suitable for detection of chromosome mutations, since large deletions and chromosomal rearrangements that extend from the transgene into chromosomal sequences will not be recovered. An exception to this is the *lacZ* plasmid assay, which will detect large deletions and chromosomal rearrangements;

however, the extent to which such mutations exhibit genetic neutrality remains to be evaluated.

Despite not being considered mandatory in the case of clear negative and positive results, molecular analysis may provide important information at several levels. First, it allows correction of mutant frequencies in those rare cases where high interanimal variability is observed or clonal expansion is suspected. Second, it allows an investigator to determine the "molecular signature" of a particular mutagen and therefore to evaluate mutational mechanisms, including the probable nature of the premutational lesion and the proximate metabolite. Third, molecular analyses allow a more precise comparison with induced mutations in endogenous targets.

Larger deletion or insertion mutations will not normally be detected by the transgenic test systems, because these alterations will not be packaged into the phage heads (Boerrigter et al., 1995). Predominantly clastogenic substances are recognized to have a low response rate in the transgenic animal mutagenicity assay, which is why the mouse micronucleus assay should be considered as a complementary assay to transgenic animal mutagenicity assays (see section 11.2 above). Nevertheless, the *gpt* delta Spi$^-$ assay (up to 10 kb) and the *lacZ* plasmid assay are able to detect large deletions. It will be important in future studies to further investigate the extent to which these assays can complement the limitation of the *lacI*, *lacZ*, *gpt* and *cII* assays.

12.5 Animal welfare and economy

Although the transgenic assays are currently more difficult to conduct and somewhat more costly than the in vivo mouse bone marrow micronucleus test, they have several advantages.

Due to considerations of animal welfare, the trend in toxicity testing is for a reduction in in vivo testing, when possible. Neither Big Blue® nor Muta™Mouse assays are animal-intensive test systems; approximately five animals per dose (three doses recommended) are recommended to be able to accurately detect a 2-fold increase in the mutation frequency (Mirsalis et al., 1995). Furthermore, the transgenic mouse assay can, in principle, be combined with other in vivo genotoxic end-points in the same animal — e.g.

micronuclei, chromosomal aberrations, UDS and sister chromatid exchange (RIVM, 2000) — and several examples in the literature demonstrate this point. Moreover, the transgenic animal models may be used to evaluate both germ and somatic cell mutagenesis in the same animals, although it would be necessary to design such a study very carefully; treatment and manifestation times have to be suitable for the detection of mutations in the two different tissues.

The availability of transgenic rat mutation models is important, since transgenic rats can be used to gather additional information from a 28-day repeated-dose toxicity assay and may also be used to conduct a 2-year carcinogenicity bioassay.

Recently, several cell lines derived from transgenic animals have been developed that allow direct comparisons between mutation in vivo and in vitro. In the future, as testing procedures attempt to replace in vivo assays when possible, the availability of the in vitro transgenic assays may, in some contexts, provide viable surrogates for in vivo gene mutation assays. Such an approach would reduce the assumptions made in comparing results obtained in vitro and in the animal.

13. CONCLUSIONS

- A positive result in a transgenic mutation assay demonstrates that the agent is a mutagen in vivo and is highly predictive of carcinogenicity.

- A negative result in a properly conducted transgenic mutation assay demonstrates that the agent is not a gene mutagen.

- Sequencing of mutations is not mandatory for assessing positive and negative results, but can provide useful mechanistic information.

- An important conclusion from this document is that the results of transgenic mutation assays complement those of the micronucleus test. Since these assays detect different types of genetic damage, this complementarity is reasonable.

- The UDS assay has often been used as a surrogate for an in vivo assay of somatic mutations. It is now possible to replace UDS with a direct assay for somatic mutation.

- Several of the transgenic assays for somatic mutation (specifically *lacI*, *lacZ*, *cII* and *gpt* delta) seem to be equivalent and can be used interchangeably with the recommended protocol. Assays like the *lacZ* plasmid assay can be used in the recommended protocol for intragenic mutations, but the protocol may not be optimal for detecting large deletions.

- Using these transgenic assays in rats is regarded as equivalent to using them in mice, and it is thought that the same recommended protocol would work equally well in both species.

14. RECOMMENDATIONS

1. It is recommended that an OECD Test Guideline be developed for these assays.

2. It is recommended that the transgenic mutation assays be included in the IPCS Qualitative Scheme for Mutagenicity and all other testing strategies.

Recommendations for future research

1. Given that the database contains few non-carcinogens, a number of well established non-carcinogens should be tested according to a robust protocol, such as that recommended by the IWGT (Thybaud et al., 2003). Such a study should include structural analogues of well established carcinogens.

2. It is recommended that these assays be used for studies of the mechanistic relationship between mutation and carcinogenesis and for studies of germline mutagenesis.

REFERENCES

ACGIH (1998) β-Propiolactone. In: Documentation of the threshold limit values (TLV) and biological exposure indices (BEI). Cincinnati, Ohio, American Conference of Governmental Industrial Hygienists, pp 1292–1293.

Adler I-D, Cao J, Filser JG, Gassner P, Kessler W, Kliesch U, Neuhäuser-Klaus A & Nüsse M (1994) Mutagenicity of 1,3-butadiene inhalation in somatic and germinal cells of mice. Mutat Res, **309**(2): 307–314.

Albanese R, Mirkova E, Gatehouse D & Ashby J (1988) Species-specific response to the rodent carcinogens 1,2-dimethylhydrazine and 1,2-dibromo-3-chloropropane in rodent bone-marrow micronucleus assays. Mutagenesis, **3**(1): 35–38.

Arai T, Kelly VP, Minowa O, Noda T & Nishimura S (2002) High accumulation of oxidative DNA damage, 8-hydroxyguanine, in *Mmh/Ogg1* deficient mice by chronic oxidative stress. Carcinogenesis, **23**: 2005–2010.

Arai T, Kelly VP, Komoro K, Minowa O, Noda T & Nishimura S (2003) Cell proliferation in liver of *Mmh/Ogg1*-deficient mice enhances mutation frequency because of the presence of 8-hydroxyguanine in DNA. Cancer Res, **63**: 4287–4292.

Araki A, Kato F, Matsushima T, Ikawa N & Nozaki K (1995) Micronuclei induction of methyl bromide in rats and mice by sub-chronic inhalation toxicity test. Environ Mutagen Res Commun, **17**(1): 47–56.

Arlt VM, Zhan L, Schmeiser HH, Honma M, Hayashi M, Phillips DH & Suzuki T (2004) DNA adducts and mutagenic specificity of the ubiquitous environmental pollutant 3-nitrobenzanthrone in Muta™Mouse. Environ Mol Mutagen, **43**(3): 186–195.

Ashby J, Brusick D, Myhr BC, Jones NJ, Parry JM, Nesnow S, Paton D, Tinwell H, Rosenkranz HS, Curti S, Gilman D & Callander RD (1993) Correlation of carcinogenic potency with mouse-skin ^{32}P-postlabeling and Muta™Mouse *lacZ*$^-$ mutation data for DMBA and its K-region sulphur isostere: comparison with activities observed in standard genotoxicity assays. Mutat Res, **292**: 25–40.

Ashby J, Short JM, Jones NJ, Lefevre PA, Provost GS, Rogers BJ, Martin EA, Parry JM, Burnette K, Glickman BW & Tinwell H (1994) Mutagenicity of *o*-anisidine to the bladder of *lacI*$^-$ transgenic B6C3F1 mice: absence of ^{14}C or ^{32}P bladder DNA adduction. Carcinogenesis, **15**(10): 2291–2296.

Ashby J, Gorelick NJ & Shelby MD (1997) Mutation assays in male germ cells from transgenic mice: overview of study and conclusions. Mutat Res, **388**: 111–122.

Asita AO, Hayashi M, Kodama Y, Matsuoka A, Suzuki T & Sofuni T (1992) Micronucleated reticulocyte induction by ethylating agents in mice. Mutat Res, **271**: 29–37.

Autrup H, Jorgensen ECB & Jensen O (1996) Aflatoxin B1 induced *lacI* mutation in liver and kidney of transgenic mice C57BL/6N: effect of phorone. Mutagenesis, **11**: 69–73.

Bastlová T, Andersson B, Lambert B & Kolman A (1993) Molecular analysis of ethylene oxide–induced mutations at the *HPRT* locus in human diploid fibroblasts. Mutat Res, **287**: 283–292.

Belitsky GA, Lytcheva TA, Khitrovo IA, Safaev RD, Zhurkov VS, Vyskubenko IF, Sytshova LP, Salamatova OG, Feldt EG, Khudoley VV, Mizgirev IV, Khovanova EM, Ugnivenko EG, Tanirbergenov TB, Malinovskaya KI, Revazova YA, Ingel FI, Bratslavsky VA, Terentyev AB, Shapiro AA & Williams GM (1994) Genotoxicity and carcinogenicity testing of 1,2-dibromopropane and 1,1,3-tribromopropane in comparison to 1,2-dibromo-3-chloropropane. Cell Biol Toxicol, **10**: 265–279.

Besaratinia A & Pfeifer GP (2003) Weak yet distinct mutagenicity of acrylamide in mammalian cells. J Natl Cancer Inst, **95**(12): 889–896.

Besaratinia A & Pfeifer GP (2004) Genotoxicity of acrylamide and glycidamide. J Natl Cancer Inst, **96**(13): 1023–1029.

Besaratinia A, Synold TW, Xi B & Pfeifer GP (2004) G-to-T transversions and small tandem base deletions are hallmark of mutations induced by ultraviolet A radiation in mammalian cells. Biochemistry (Washington), **43**(25): 8169–8177.

Bielas JH (2002) A more efficient Big Blue® protocol improves transgene rescue and accuracy in an adduct and mutation measurement. Mutat Res, **518**(2): 107–112.

Bielas JH & Heddle JA (2000) Proliferation is necessary for both repair and mutation in transgenic mouse cells. Proc Natl Acad Sci USA, **97**(21): 11391–11396.

Blakey DH, Douglas GR, Huang KC & Winter HI (1995) Cytogenetic mapping of lambda gt10 *lacZ* sequences in transgenic mouse strain 40.6 (Muta™Mouse). Mutagenesis, **10**: 145–148.

Boerrigter ME (1999) Treatment of *lacZ* plasmid-based transgenic mice with benzo[*a*]pyrene: measurement of DNA adduct levels, mutant frequencies, and mutant spectra. Environ Mol Mutagen, **34**: 140–147.

Boerrigter METI, Dollé MET, Martus H-J, Gossen JA & Vijg J (1995) Plasmid-based transgenic mouse model for studying in vivo mutations. Nature, **377**: 657–659.

Bol SA, Horlbeck J, Markovic J, de Boer JG, Turesky RJ & Constable A (2000) Mutational analysis of the liver, colon and kidney of Big Blue® rats treated with 2-amino-3-methylimidazo[4,5-*f*]quinoline. Carcinogenesis, **21**: 1–6.

Boyiri T, Guttenplan J, Khmelnitsky M, Kosinska W, Lin JM, Desai D, Amin S, Pittman B & El-Bayoumy K (2004) Mammary carcinogenesis and molecular analysis of in vivo *cII* gene mutations in the mammary tissue of female transgenic rats treated with the environmental pollutant 6-nitrochrysene. Carcinogenesis, **25**(4): 637–643.

Brault D, Bouilly C, Renault D & Thybaud V (1996) Tissue-specific induction of mutations by acute oral administration of *N*-methyl-*N*′-nitro-*N*-nitrosoguanidine and β-propiolactone to the Muta™Mouse: preliminary data on stomach, liver and bone marrow. Mutat Res, **360**: 83–87.

Brooks TM & Dean SW (1996) Detection of gene mutation in skin, stomach and liver of Muta™Mouse following oral or topical treatment with *N*-methyl-*N*'-nitro-*N*-nitrosoguanidine or 1-chloromethylpyrene: some preliminary observations. Mutagenesis, **11**: 529–532.

Brooks TM & Dean SW (1997) The detection of gene mutation in the tubular sperm of Muta™Mice following a single intraperitoneal treatment with methyl methanesulphonate or ethylnitrosourea. Mutat Res, **388**: 219–222.

Brooks TM, Szegedi M, Rosher P & Dean SW (1995) The detection of gene mutation in transgenic mice (Muta™Mouse) following a single oral dose of 2-acetylaminofluorene. Mutagenesis, **10**: 149–150.

BUA (1996) Hydrazin, Hydrazinhydrat und Hydrazinsulfate. Beratergremium für umweltrelevante Altstoffe. Stuttgart, S. Hirzel Wissenschaftliche Verlagsgesellschaft, pp 1–179 (BUA Report No. 205).

Buettner VL, Nishino H, Haavik J, Knoll A, Hill K & Sommer SS (1997) Spontaneous mutation frequencies and spectra in p53(+/+) and p53(−/−) mice: a test of the "guardian of the genome" hypothesis in the Big Blue® transgenic mouse mutation detection system. Mutat Res, **379**(1): 13–20.

Buettner VL, Hill KA, Halangoda A & Sommer SS (1999) Tandem-base mutations occur in mouse liver and adipose tissue preferentially as G:C→T:A transversions and accumulate with age. Environ Mol Mutagen, **33**(4): 320–324.

Burkhart JG, Burkhart BA, Sampson KS & Malling HV (1993) ENU-induced mutagenesis at a single A-T base pair in transgenic mice containing Phi-X174. Mutat Res, **292**: 69–81.

Butterworth BE, Templin MV, Constan AA, Sprankle CS, Wong BA, Pluta LJ, Everitt JI & Recio L (1998) Long-term mutagenicity studies with chloroform and dimethylnitrosamine in female *lacI* transgenic B6C3F1 mice. Environ Mol Mutagen, **31**: 248–256.

Cariello NF, Douglas GR, Gorelick NJ, Hart DW, Wilson JD & Soussi T (1998) Databases and software for the analysis of mutations in the human *p53* gene, human *hprt* gene and both the *lacI* and *lacZ* gene in transgenic rodents. Nucl Acids Res, **26**: 198–199.

Carr GJ & Gorelick NJ (1994) Statistical tests of significance in transgenic mutation assays: considerations on the experimental unit. Environ Mol Mutagen, **24**(4): 276–282.

Carr GJ & Gorelick NJ (1995) Statistical design and analysis of mutation studies in transgenic mice. Environ Mol Mutagen, **25**(3): 246–255.

Carr GJ & Gorelick NJ (1996) A place for statistics in the generation and analysis of genetic toxicity data: a response to "rodent mutation assay data presentation and statistical assessment." Mutat Res, **357**(1–2): 257–260.

Casciano DA, Aidoo A, Chen T, Mittelstaedt RA, Manjanatha MG & Heflich RH (1999) *Hprt* mutant frequency and molecular analysis of *Hprt* mutations in rats treated with mutagenic carcinogens. Mutat Res, **431**: 389–395.

CCRIS (1995a) 4-Acetylaminofluorene. CAS: 28322-02-3. Bethesda, Maryland, United States Department of Health and Human Services, National Institutes of Health, National Library of Medicine, Chemical Carcinogenesis Research Information System (http://toxnet.nlm.nih.gov/cgi-bin/sis/htmlgen?CCRIS).

CCRIS (1995b) Beta-Propiolactone. CAS: 57-57-8. Bethesda, Maryland, United States Department of Health and Human Services, National Institutes of Health, National Library of Medicine, Chemical Carcinogenesis Research Information System (http://toxnet.nlm.nih.gov/cgi-bin/sis/htmlgen?CCRIS).

CCRIS (1995c) Chlorambucil. CAS: 305-03-3. Bethesda, Maryland, United States Department of Health and Human Services, National Institutes of Health, National Library of Medicine, Chemical Carcinogenesis Research Information System (http://toxnet.nlm.nih.gov/cgi-bin/sis/htmlgen?CCRIS).

CCRIS (1996) Acrylamide. CAS: 79-06-1. Bethesda, Maryland, United States Department of Health and Human Services, National Institutes of Health, National Library of Medicine, Chemical Carcinogenesis Research Information System (http://toxnet.nlm.nih.gov/cgi-bin/sis/htmlgen?CCRIS).

CCRIS (2000) *N*-Propyl-*N*-nitrosourea. CAS: 816-57-9. Bethesda, Maryland, United States Department of Health and Human Services, National Institutes of Health, National Library of Medicine, Chemical Carcinogenesis Research Information System (http://toxnet.nlm.nih.gov/cgi-bin/sis/htmlgen?CCRIS).

CCRIS (2001a) Ethyl methanesulfonate. CAS: 62-50-0. Bethesda, Maryland, United States Department of Health and Human Services, National Institutes of Health, National Library of Medicine, Chemical Carcinogenesis Research Information System (http://toxnet.nlm.nih.gov/cgi-bin/sis/htmlgen?CCRIS).

CCRIS (2001b) Methyl bromide. CAS: 74-83-9. Bethesda, Maryland, United States Department of Health and Human Services, National Institutes of Health, National Library of Medicine, Chemical Carcinogenesis Research Information System (http://toxnet.nlm.nih.gov/cgi-bin/sis/htmlgen?CCRIS).

CCRIS (2001c) Tris(2,3-Dibromopropyl)phosphate. CAS: 126-72-7. Bethesda, Maryland, United States Department of Health and Human Services, National Institutes of Health, National Library of Medicine, Chemical Carcinogenesis Research Information System (http://toxnet.nlm.nih.gov/cgi-bin/sis/htmlgen?CCRIS).

CCRIS (2001d) Tamoxifen. CAS: 10540-29-1. Bethesda, Maryland, United States Department of Health and Human Services, National Institutes of Health, National Library of Medicine, Chemical Carcinogenesis Research Information System (http://toxnet.nlm.nih.gov/cgi-bin/sis/htmlgen?CCRIS).

CCRIS (2002a) 7,12-Dimethylbenz[*a*]anthracene. CAS: 57-97-6. Bethesda, Maryland, United States Department of Health and Human Services, National Institutes of Health, National Library of Medicine, Chemical Carcinogenesis Research Information System (http://toxnet.nlm.nih.gov/cgi-bin/sis/htmlgen?CCRIS).

CCRIS (2002b) Mitomycin C. CAS: 50-07-7. Bethesda, Maryland, United States Department of Health and Human Services, National Institutes of Health, National Library of Medicine, Chemical Carcinogenesis Research Information System (http://toxnet.nlm.nih.gov/cgi-bin/sis/htmlgen?CCRIS).

CCRIS (2002c) *N,N*-Dipropylnitrosamine. CAS: 621-64-7. Bethesda, Maryland, United States Department of Health and Human Services, National Institutes of Health, National Library of Medicine, Chemical Carcinogenesis Research Information System (http://toxnet.nlm.nih.gov/cgi-bin/sis/htmlgen?CCRIS).

CCRIS (2003a) 4-Aminobiphenyl. CAS: 92-67-1. Bethesda, Maryland, United States Department of Health and Human Services, National Institutes of Health, National Library of Medicine, Chemical Carcinogenesis Research Information System (http://toxnet.nlm.nih.gov/cgi-bin/sis/htmlgen?CCRIS).

CCRIS (2003b) 4-Nitroquinoline 1-oxide. CAS: 56-57-5. Bethesda, Maryland, United States Department of Health and Human Services, National Institutes of Health, National Library of Medicine, Chemical Carcinogenesis Research Information System (http://toxnet.nlm.nih.gov/cgi-bin/sis/htmlgen?CCRIS).

CCRIS (2003c) Benzo[*a*]pyrene. CAS: 50-32-8. Bethesda, Maryland, United States Department of Health and Human Services, National Institutes of Health, National Library of Medicine, Chemical Carcinogenesis Research Information System (http://toxnet.nlm.nih.gov/cgi-bin/sis/htmlgen?CCRIS).

CCRIS (2003d) *N,N*-Diethylnitrosamine. CAS: 55-18-5. Bethesda, Maryland, United States Department of Health and Human Services, National Institutes of Health, National Library of Medicine, Chemical Carcinogenesis Research Information System (http://toxnet.nlm.nih.gov/cgi-bin/sis/htmlgen?CCRIS).

CCRIS (2003e) Quinoline. CAS: 91-22-5. Bethesda, Maryland, United States Department of Health and Human Services, National Institutes of Health, National Library of Medicine, Chemical Carcinogenesis Research Information System (http://toxnet.nlm.nih.gov/cgi-bin/sis/htmlgen?CCRIS).

CCRIS (2003f) 4-(*N*-Methyl-*N*-nitrosamino)-1-(3-pyridyl)-1-butanone. CAS: 64091-91-4. Bethesda, Maryland, United States Department of Health and Human Services, National Institutes of Health, National Library of Medicine, Chemical Carcinogenesis Research Information System (http://toxnet.nlm.nih.gov/cgi-bin/sis/htmlgen?CCRIS).

Chang PY, Mirsalis J, Riccio ES, Bakke JP, Lee PS, Shimon J, Phillips S, Fairchild D, Hara Y & Crowell JA (2003) Genotoxicity and toxicity of the potential cancer-preventive agent polyphenon E. Environ Mol Mutagen, **41**: 43–54.

Chen JB, Dobrovolsky VN & Heflich RH (1999) Development of a mouse cell line containing the ΦX174 *am3* allele as a target for detecting mutation. Mutat Res, **444**: 347–353.

Chen T, Aidoo A, Manjanatha MG, Mittelstaedt RA, Shelton SD, Lyn-Cook LE, Casciano DA & Heflich RH (1998) Comparison of mutant frequencies and types of mutations induced by thiotepa in the endogenous *Hprt* gene and transgenic *lacI* gene of Big Blue® rats. Mutat Res, **403**: 199–214.

Chen T, Mittelstaedt RA, Aidoo A, Hamilton LP, Beland FA, Casciano DA & Heflich RH (2001a) Comparison of *Hprt* and *lacI* mutant frequency with DNA adduct formation in

N-hydroxy-2-acetylaminofluorene-treated Big Blue® rats. Environ Mol Mutagen, **37**(3): 195–202.

Chen T, Mittelstaedt RA, Shelton SD, Balachandra Dass S, Manjanatha MG, Casciano DA & Heflich RH (2001b) Gene- and tissue-specificity of mutation in Big Blue rats treated with the hepatocarcinogen *N*-hydroxy-2-acetylaminofluorene. Environ Mol Mutagen, **37**(3): 203–214.

Chen T, Gamboa da Costa G, Marques MM, Shelton SD, Beland FA & Manjanatha MG (2002) Mutations induced by α-hydroxytamoxifen in the *lacI* and *cII* genes of Big Blue® transgenic rats. Carcinogenesis, **23**(10): 1751–1757.

Cole J & Skopek TR (1994) Somatic mutant frequency, mutation rates and mutational spectra in the human population in vivo. Mutat Res, **304**: 33–105.

Cosentino L & Heddle JA (1999) A comparison of the effects of diverse mutagens at the *lacZ* transgene and *Dlb-1* locus in vivo. Mutagenesis, **14**: 113–119.

Cosentino L & Heddle JA (2000) Differential mutation of transgenic and endogenous loci in vivo. Mutat Res, **454**(1–2): 1–10.

Cunningham MJ, Choy WN, Arce GT, Rickard LB, Vlachos DA, Kinney LA & Sarrif AM (1986) In vivo sister chromatid exchange and micronucleus induction studies with 1,3-butadiene in B6C3F1 mice and Sprague-Dawley rats. Mutagenesis, **1**: 449–452.

Cunningham ML, Hayward JJ, Shane BS & Tindall KR (1996) Distinction of mutagenic carcinogens from a mutagenic noncarcinogen in the Big Blue® transgenic mouse. Environ Health Perspect, **104**: 683–686.

da Costa GG, Manjanatha MG, Marques MM & Beland FA (2002) Induction of *lacI* mutations in Big Blue® rats treated with tamoxifen and α-hydroxytamoxifen. Cancer Lett, **176**(1): 37–45.

Davies R, Oreffo VIC, Martin EA, Festing MFW, White INH, Smith LL & Styles JA (1997) Tamoxifen causes gene mutations in the livers of lambda/*lacI* transgenic rats. Cancer Res, **57**: 1288–1293.

Davies R, Gant TW, Smith LL & Styles JA (1999) Tamoxifen induces G:C→T:A mutations in the *cII* gene in the liver of λ/*lacI* transgenic rats but not at 5'-CpG-3' dinucleotide sequences as found in the *lacI* transgene. Carcinogenesis, **20**: 1351–1356.

Davis CD, Dacquel EJ, Schut HAJ, Thorgeirsson SS & Snyderwine EG (1996) In vivo mutagenicity and DNA adduct levels of heterocyclic amines in Muta™Mice and *c-myc*/*lacZ* double transgenic mice. Mutat Res, **356**: 287–296.

Dean SW & Myhr B (1994) Measurement of gene mutation in vivo using Muta™Mouse and positive selection for *lacZ*⁻ phage. Mutagenesis, **9**: 183–185.

Dean SW, Brooks TM, Burlinson B, Mirsalis J, Myhr B, Recio L & Thybaud V (1999) Transgenic mouse mutation assay systems can play an important role in regulatory mutagenicity testing in vivo for the detection of site-of-contact mutagens. Mutagenesis, **14**(1): 141–151.

de Boer JG & Glickman BW (1998) The *lacI* gene as a target for mutation in transgenic rodents and *Escherichia coli*. Genetics, **148**: 1441–1451.

de Boer JG, Mirsalis JC, Provost GS, Tindall KR & Glickman BW (1996) Spectrum of mutations in kidney, stomach, and liver from *lacI* transgenic mice recovered after treatment with tris(2,3-dibromopropyl)phosphate. Environ Mol Mutagen, **28**: 418–423.

de Boer JG, Provost S, Gorelick N, Tindal K & Glickman BW (1998) Spontaneous mutation in *lacI* transgenic mice: a comparison of tissues. Mutagenesis, **13**: 109–114.

DECOS (1989) Health-based recommended occupational exposure limits for ethyl methanesulphonate (EMS) methyl methanesulphonate (MMS). Voorburg, Directorate-General of Labour, Dutch Expert Committee for Occupational Standards, pp 1–73 (RA 4/89).

Delongchamp RR, Malling HV, Chen JB & Heflich RH (1999) An estimator of the mutant frequency in assays using transgenic animals. Mutat Res, **440**: 101–108.

Delongchamp RR, Valentine CR & Malling HV (2001) Estimation of the average burst size of ΦX174 *am3*, *cs70* for use in mutation assays with transgenic mice. Environ Mol Mutagen, **37**: 356–360.

De Silva IU, McHugh PJ, Clingen PH & Hartley JA (2000) Defining the roles of nucleotide excision repair and recombination in the repair of DNA interstrand cross-links in mammalian cells. Mol Cell Biol, **20**(21): 7980–7990.

Dobrovolsky VN, Casciano DA & Heflich RH (1999) *Tk+/−* mouse model for detecting in vivo mutation in an endogenous, autosomal gene. Mutat Res, **423**: 125–136.

Dobrovolsky VN, Shaddock JG & Heflich RH (2005) Analysis of in vivo mutation in the *Hprt* and *Tk* genes of mouse lymphocytes. In: Keohavong P & Grant SG eds. Methods in molecular biology. Vol. 291. Molecular toxicology protocols. Totowa, New Jersey, Humana Press Inc., pp 133–144.

Dollé ME & Vijg J (2002) Genome dynamics in aging mice. Genome Res, **12**: 1732–1738.

Dollé ME, Martus HJ, Gossen JA, Boerrigter ME & Vijg J (1996) Evaluation of a plasmid-based transgenic mouse model for detecting in vivo mutations. Mutagenesis, **11**: 111–118.

Dollé ME, Giese H, Hopkins CL, Martus HJ, Hausdorff JM & Vijg J (1997) Rapid accumulation of genome rearrangements in liver but not in brain of old mice. Nat Genet, **17**: 431–434.

Dollé ME, Snyder WK, Gossen JA, Lohman PHM & Vijg J (2000) Distinct spectra of somatic mutations accumulated with age in mouse heart and small intestine. Proc Natl Acad Sci USA, **97**: 8403–8408.

Dollé ME, Snyder WK, Dunson DB & Vijg J (2002) Mutational fingerprints of aging. Nucleic Acids Res, **30**: 545–549.

Douglas GR, Gingerich JD, Gossen JA & Bartlett SA (1994) Sequence spectra of spontaneous *lacZ* gene mutations in transgenic mouse somatic and germline tissues. Mutagenesis, **9**: 451–458.

Douglas GR, Jiao J, Gingerich JD, Gossen JA & Soper LM (1995a) Temporal and molecular characteristics of mutations induced by ethylnitrosourea in germ cells isolated from seminiferous tubules and in spermatozoa of *lacZ* transgenic mice. Proc Natl Acad Sci USA, **92**: 7485–7489.

Douglas GR, Gingerich JD & Soper LM (1995b) Evidence for in vivo non-mutagenicity of the carcinogen hydrazine sulfate in target tissues of *lacZ* transgenic mice. Carcinogenesis, **16**(4): 801–804.

Douglas GR, Jiao J, Gingerich JD, Soper LM & Gossen JA (1996) Temporal and molecular characteristics of *lacZ* mutations in somatic tissues of transgenic mice. Environ Mol Mutagen, **28**: 317–324.

Douglas GR, Gingerich JD, Soper LM, Potvin M & Bjarnason S (1999) Evidence for the lack of base-change and small-deletion mutation induction by trichloroethylene in *lacZ* transgenic mice. Environ Mol Mutagen, **34**: 190–194.

Dybdahl M, Risom L, Bornholdt J, Autrup H, Loft S & Wallin H (2004) Inflammatory and genotoxic effects of diesel particles in vitro and in vivo. Mutat Res, **562**(1–2): 119–131.

Dycaico MJ, Provost GS, Kretz PL, Ransom SL, Moores JC & Short JM (1994) The use of shuttle vectors for mutation analysis in transgenic mice and rats. Mutat Res, **307**(2): 461–478.

Dycaico MJ, Stuart GR, Tobal GM, de Boer JG, Glickman BW & Provost GS (1996) Species-specific differences in hepatic mutant frequency and mutational spectrum among λ/*lacI* transgenic rats and mice following exposure to aflatoxin B1. Carcinogenesis, **17**: 2347–2356.

Eckhardt K, King M-T, Gocke E & Wild D (1980) Mutagenicity study of Remsen-Fahlberg saccharin and contaminants. Toxicol Lett, **7**(1): 51–60.

Engle SJ, Stockelman MG, Chen J, Boivin G, Yum MN, Davies PM, Ying MY, Sahota A, Simmonds HA, Stambrook P & Tischfield JA (1996) Adenine phosphoribosyltransferase-deficient mice develop 2,8-dihydroxyadenine nephrolithiasis. Proc Natl Acad Sci USA, **93**: 5307–5312.

Erexson GL, Watson DE & Tindall KR (1999) Characterization of new transgenic Big Blue® mouse and rat primary fibroblast cell strains for use in molecular toxicology studies. Environ Mol Mutagen, **34**: 90–96.

European Chemicals Bureau (2002) European Union risk assessment report: *o*-Anisidine. Luxembourg, European Chemicals Bureau, Institute for Health and Consumer Protection, Office for Official Publications of the European Communities, pp 1–94.

Fahrig R (1977) The mammalian spot test (Fellfleckentest) with mice. Arch Toxicol, **38**: 87–98.

Fahrig R (1988) Positive response of 2-acetylaminofluorene, negative response of 4-acetylaminofluorene in the mammalian spot test. In: Ashby J ed. Evaluation of short-term tests for carcinogens: report of the International Programme on Chemical Safety's collaborative study on in vivo assays. Cambridge, Cambridge University Press, pp 159–163.

Fahrig R (1993) Genetic effects of dioxins in the spot test with mice. Environ Health Perspect, **101**(Suppl. 3): 257–261.

Fahrig R & Steinkamp-Zucht A (1996) Co-recombinogenic and anti-mutagenic effects of diethylhexylphthalate, inactiveness of pentachlorophenol in the spot test with mice. Mutat Res, **354**: 59–67.

Farabaugh PJ (1978) Sequence of the *lacI* gene. Nature, **274**(5673): 765–769.

Fletcher K, Tinwell H & Ashby J (1998) Mutagenicity of the human bladder carcinogen 4-aminobiphenyl to the bladder of Muta™Mouse transgenic mice. Mutat Res, **400**: 245–250.

Fletcher K, Soames AR, Tinwell H, Lefevre PA & Ashby J (1999) Hepatic gene mutations induced in Big Blue® rats by both the potent rat liver azo-carcinogen 6BT and its reported noncarcinogenic analogue 5BT. Environ Mol Mutagen, **34**: 148–153.

Friedberg EC, Walker GC & Siede W ed (1995) DNA repair and mutagenesis. Washington, DC, ASM Press, pp 1–698.

Furuno-Fukushi I, Masumura K, Furuse T, Noda Y, Takahagi M, Saito T, Hoki Y, Suzuki H, Wynshaw-Boris A, Nohmi T & Tatsumi K (2003) Effect of Atm disruption on spontaneously arising and radiation-induced deletion mutations in mouse liver. Radiat Res, **160**: 549–558.

GENE-TOX (1992) *N*-Propyl-*N*-nitrosourea. CAS: 816-57-9. Bethesda, Maryland, United States Department of Health and Human Services, National Institutes of Health, National Library of Medicine, Genetic Toxicology (Mutagenicity) (http://toxnet.nlm.nih.gov/cgi-bin/sis/htmlgen?GENETOX).

GENE-TOX (1995a) Beta-Propiolactone. CAS: 57-57-8. Bethesda, Maryland, United States Department of Health and Human Services, National Institutes of Health, National Library of Medicine, Genetic Toxicology (Mutagenicity) (http://toxnet.nlm.nih.gov/cgi-bin/sis/htmlgen?GENETOX).

GENE-TOX (1995b) Dimethylnitrosamine. CAS: 62-75-9. Bethesda, Maryland, United States Department of Health and Human Services, National Institutes of Health, National Library of Medicine, Genetic Toxicology (Mutagenicity) (http://toxnet.nlm.nih.gov/cgi-bin/sis/htmlgen?GENETOX).

GENE-TOX (1995c) Ethyl methanesulfonate. CAS: 62-50-0. Bethesda, Maryland, United States Department of Health and Human Services, National Institutes of Health, National Library of Medicine, Genetic Toxicology (Mutagenicity) (http://toxnet.nlm.nih.gov/cgi-bin/sis/htmlgen?GENETOX).

GENE-TOX (1995d) Quinoline. CAS: 91-22-5. Bethesda, Maryland, United States Department of Health and Human Services, National Institutes of Health, National Library of Medicine, Genetic Toxicology (Mutagenicity) (http://toxnet.nlm.nih.gov/cgi-bin/sis/htmlgen?GENETOX).

GENE-TOX (1998a) 1-Ethyl-1-nitrosourea. CAS: 759-73-9. Bethesda, Maryland, United States Department of Health and Human Services, National Institutes of Health, National Library of Medicine, Genetic Toxicology (Mutagenicity) (http://toxnet.nlm.nih.gov/cgi-bin/sis/htmlgen?GENETOX).

GENE-TOX (1998b) 1-Methyl-1-nitrosourea. CAS: 684-93-5. Bethesda, Maryland, United States Department of Health and Human Services, National Institutes of Health, National Library of Medicine, Genetic Toxicology (Mutagenicity) (http://toxnet.nlm.nih.gov/cgi-bin/sis/htmlgen?GENETOX).

GENE-TOX (1998c) 4-Acetylaminofluorene. CAS: 28322-02-3. Bethesda, Maryland, United States Department of Health and Human Services, National Institutes of Health, National Library of Medicine, Genetic Toxicology (Mutagenicity) (http://toxnet.nlm.nih.gov/cgi-bin/sis/htmlgen?GENETOX).

GENE-TOX (1998d) 4-Amino-1,1'-biphenyl. CAS: 92-67-1. Bethesda, Maryland, United States Department of Health and Human Services, National Institutes of Health, National Library of Medicine, Genetic Toxicology (Mutagenicity) (http://toxnet.nlm.nih.gov/cgi-bin/sis/htmlgen?GENETOX).

GENE-TOX (1998e) 4-Nitroquinoline 1-oxide. CAS: 56-57-5. Bethesda, Maryland, United States Department of Health and Human Services, National Institutes of Health, National Library of Medicine, Genetic Toxicology (Mutagenicity) (http://toxnet.nlm.nih.gov/cgi-bin/sis/htmlgen?GENETOX).

GENE-TOX (1998f) 7,12-Dimethylbenz[*a*]anthracene. CAS: 57-97-6. Bethesda, Maryland, United States Department of Health and Human Services, National Institutes of Health, National Library of Medicine, Genetic Toxicology (Mutagenicity) (http://toxnet.nlm.nih.gov/cgi-bin/sis/htmlgen?GENETOX).

GENE-TOX (1998g) Chlorambucil. CAS: 305-03-3. Bethesda, Maryland, United States Department of Health and Human Services, National Institutes of Health, National Library of Medicine, Genetic Toxicology (Mutagenicity) (http://toxnet.nlm.nih.gov/cgi-bin/sis/htmlgen?GENETOX).

GENE-TOX (1998h) Diethylnitrosamine. CAS: 55-18-5. Bethesda, Maryland, United States Department of Health and Human Services, National Institutes of Health, National Library of Medicine, Genetic Toxicology (Mutagenicity) (http://toxnet.nlm.nih.gov/cgi-bin/sis/htmlgen?GENETOX).

GENE-TOX (1998i) Mitomycin C. CAS: 50-07-7. Bethesda, Maryland, United States Department of Health and Human Services, National Institutes of Health, National Library of Medicine, Genetic Toxicology (Mutagenicity) (http://toxnet.nlm.nih.gov/cgi-bin/sis/htmlgen?GENETOX).

GENE-TOX (1998j) *N,N*-Dipropylnitrosamine. CAS: 621-64-7. Bethesda, Maryland, United States Department of Health and Human Services, National Institutes of Health, National Library of Medicine, Genetic Toxicology (Mutagenicity) (http://toxnet.nlm.nih.gov/cgi-bin/sis/htmlgen?GENETOX).

GENE-TOX (1998k) Tris(2,3-Dibromopropyl)phosphate. CAS: 126-72-7. Bethesda, Maryland, United States Department of Health and Human Services, National Institutes of Health, National Library of Medicine, Genetic Toxicology (Mutagenicity) (http://toxnet.nlm.nih.gov/cgi-bin/sis/htmlgen?GENETOX).

GENE-TOX (1998l) Urethane. CAS: 51-79-6. Bethesda, Maryland, United States Department of Health and Human Services, National Institutes of Health, National Library of Medicine, Genetic Toxicology (Mutagenicity) (http://toxnet.nlm.nih.gov/cgi-bin/sis/htmlgen?GENETOX).

Gilbert W & Müller-Hill B (1967) The *lac* operator is DNA. Proc Natl Acad Sci USA, **58**: 2415–2421.

Glazer PM, Sarkar SN & Summers WC (1986) Detection and analysis of UV-induced mutations in mammalian cell DNA using a lambda phage shuttle vector. Proc Natl Acad Sci USA, **83**: 1041–1044.

Gocke E, Wild D, Eckhardt K & King MT (1983) Mutagenicity studies with the mouse spot test. Mutat Res, **117**: 201–212.

Gollapudi BB, Jackson KM & Stott WT (1998) Hepatic *lacI* and *cII* mutation in transgenic (λLIZ) rats treated with dimethylnitrosamine. Mutat Res, **419**: 131–135.

Gondo Y, Shioyama Y, Nakao K & Katsuki M (1996) A novel positive detection system of in vivo mutations in *rpsL* (*strA*) transgenic mice. Mutat Res, **360**: 1–14.

Gordon JW & Ruddle FH (1983) Gene transfer into mouse embryos: production of transgenic mice by pronuclear injection. Meth Enzymol, **101**: 411–433.

Gordon JW, Scangos GA, Plotkin DJ, Barbosa JA & Ruddle FH (1980) Genetic transformation of mouse embryos by microinjection of purified DNA. Proc Natl Acad Sci USA, **77**(12): 7380–7384.

Gorelick N (1995) Overview of mutation assays in transgenic mice for routine testing. Environ Mol Mutagen, **25**: 218–230.

Gorelick NJ & Thompson ED (1994) Overview of the workshops on statistical analysis of mutation data from transgenic mice. Environ Mol Mutagen, **23**: 12–16.

Gorelick NJ, Andrews JL, Gu M & Glickman BW (1995) Mutational spectra in the *lacI* gene in skin from 7,12-dimethylbenz[*a*]anthracene-treated and untreated transgenic mice. Mol Carcinog, **14**: 53–62.

Gorelick NJ, Andrews JL, Gibson DP, Carr GJ & Aardema MJ (1997) Evaluation of *lacI* mutation in germ cells and micronuclei in peripheral blood after treatment of male *lacI* transgenic mice with ethylnitrosourea, isopropylmethane sulfonate or methylmethane sulfonate. Mutat Res, **388**: 187–195.

Gorelick NJ, Andrews JL, de Boer JG, Young R, Gibson DP & Walker VE (1999) Tissue-specific mutant frequencies and mutational spectra in cyclophosphamide-treated *lacI* transgenic mice. Environ Mol Mutagen, **34**: 154–166.

Gossen J & Vijg J (1993) Transgenic mice as model systems for studying gene mutations in vivo. Trends Genet, **9**: 27–31.

Gossen JA, De Leeuw WJF, Tan CHT, Zwarthoff EC, Berends F, Lohman PHM, Knook DL & Vijg J (1989) Efficient rescue of integrated shuttle vectors from transgenic mice: a model for studying mutations in vivo. Proc Natl Acad Sci USA, **86**: 7971–7975.

Gossen JA, Molijn AC, Douglas GR & Vijg J (1992) Application of galactose-sensitive *E. coli* strains as selective hosts for *lacZ*$^-$ plasmids. Nucleic Acids Res, **20**: 3254.

Gossen JA, de Leeuw WJF, Molijn AC & Vijg J (1993) Plasmid rescue from transgenic mouse DNA using LacI repressor protein conjugated to magnetic beads. BioTechniques, **14**(4): 624–629.

Gossen JA, Martus H-J, Wei JY & Vijg J (1995) Spontaneous and X-ray-induced deletion mutations in a LacZ plasmid-based transgenic mouse model. Mutat Res, **331**: 89–97.

Gunz D, Shephard SE & Lutz WK (1993) Can nongenotoxic carcinogens be detected with the *lacI* transgenic mouse mutation assay? Environ Mol Mutagen, **21**: 209–211.

Hachiya N & Motohashi Y (2000) Examination of *lacZ* mutant induction in the liver and testis of Muta Mouse following injection of halogenated aliphatic hydrocarbons classified as human carcinogens. Ind Health, **38**(2): 213–220.

Hachiya N, Yajima N, Hatakeyama S, Yuno K, Okada N, Umeda Y, Wakata A & Motohashi Y (1999) Induction of *lacZ* mutation by 7,12-dimethylbenz[*a*]anthracene in various tissues of transgenic mice. Mutat Res, **444**: 283–295.

Hakura A, Tsutsui Y, Sonoda J, Kai J, Imade T, Shimada M, Sugihara Y & Mikami T (1998) Comparison between in vivo mutagenicity and carcinogenicity in multiple organs by benzo[*a*]pyrene in the *lacZ* transgenic mouse (Muta™Mouse). Mutat Res, **398**: 123–130.

Hakura A, Tsutsui Y, Sonoda J, Mikami T, Tsukidate K, Sagami F & Kerns WD (1999) Multiple organ mutation in the *lacZ* transgenic mouse (Muta™Mouse) 6 months after oral treatment (5 days) with benzo[*a*]pyrene. Mutat Res, **426**(1): 71–77.

Hakura A, Tsutsui Y, Sonoda J, Tsukidate K, Mikami T & Sagami F (2000) Comparison of the mutational spectra of the *lacZ* transgene in four organs of the Muta™Mouse treated with benzo[*a*]pyrene: target organ specificity. Mutat Res, **447**(2): 239–247.

Halangoda A, Still JG, Hill KA & Sommer SS (2001) Spontaneous microdeletions and microinsertions in a transgenic mouse mutation detection system: analysis of age, tissue, and sequence specificity. Environ Mol Mutagen, **37**: 311–323.

Hamoud MA, Ong T, Petersen M & Nath J (1989) Effects of quinoline and 8-hydroxyquinoline on mouse bone marrow erythrocytes as measured by the micronucleus assay. Teratog Carcinog Mutagen, **9**: 111–118.

Hansen M, Hald MT, Autrup H, Vogel U, Bornholdt J, Moller P, Molck AM, Lindecrona R, Poulsen HE, Wallin H, Loft S & Dragsted LO (2004) Sucrose and IQ induced mutations in rat colon by independent mechanism. Mutat Res, **554**(1–2): 279–286.

Hara T, Hirano K, Hirano N, Tamura H, Sui H, Shibuya T, Hyogo A, Hirashio T, Tokai H, Yamashita Y & Kura K (1999) Mutation induction by *N*-propyl-*N*-nitrosourea in eight Muta™Mouse organs. Mutat Res, **444**: 297–307.

Harbach PR, Zimmer DM, Filipunas AL, Mattes WB & Aaron CS (1999) Spontaneous mutation spectrum at the lambda *cII* locus in liver, lung, and spleen tissue of Big Blue® transgenic mice. Environ Mol Mutagen, **33**(2): 132–143.

Hart J (1985) The mouse spot test: results with a new cross. Arch Toxicol, **58**(1): 1–4.

Hashimoto K, Ohshawa K & Kimura M (2004) Mutations induced by 4-(methylnitrosamino)-1-(3-pyridyl)-1-butanone (NNK) in the *lacZ* and *cII* genes of Muta™Mouse. Mutat Res, **560**(2): 119–131.

Hayashi H, Kondo H, Masumura K, Shindo Y & Nohmi T (2003) Novel transgenic rat for in vivo genotoxic assays using 6-thioguanine and Spi⁻ selection. Environ Mol Mutagen, **41**: 253–259.

Hayward JJ, Shane BS, Tindall KR & Cunningham ML (1995) Differential in vivo mutagenicity of the carcinogen/noncarcinogen pair 2,4- and 2,6-diaminotoluene. Carcinogenesis, **16**(10): 2429–2433.

Health Canada (2004) Detailed review of transgenic rodent mutation assays (draft). Ottawa, Ontario, Health Canada, pp 1–584.

Heddle JA (1999) Mutant manifestation: the time factor in somatic mutagenesis. Mutagenesis, **14**: 1–3.

Heddle JA, Hite M, Kirkhart B, Mavournin K, MacGregor JT, Newell GW & Salamone MF (1983) The induction of micronuclei as a measure of genotoxicity: A report of the U.S. Environmental Protection Agency GENE-TOX Program. Mutat Res, **123**(1): 61–118.

Heddle JA, Dean S, Nohmi T, Boerrigter M, Casciano D, Douglas GR, Glickman BW, Gorelick NJ, Mirsalis JC, Martus H-J, Skopek TR, Thybaud V, Tindall KR & Yajima N (2000) In vivo transgenic mutation assays. Environ Mol Mutagen, **35**: 253–259.

Heddle JA, Martus HJ & Douglas GR (2003) Treatment and sampling protocols for transgenic mutation assays. Environ Mol Mutagen, **41**(1): 1–6.

Hill KA, Wang J, Farwell KD & Sommer SS (2003) Spontaneous tandem-base mutations (TBM) show dramatic tissue, age, pattern and spectrum specificity. Mutat Res, **534**: 173–183.

Hill KA, Buettner VL, Halangoda A, Kunishige M, Moore SR, Scaringe WA & Sommer SS (2004) Spontaneous mutation in Big Blue® mice from fetus to old age: Tissue-specific time courses of mutation frequency but similar mutation types. Environ Mol Mutagen, **43**(2): 110–120.

Hoogervorst EM, van Oostrom CT, Beems RB, van Benthem J, Gielis S, Vermeulen JP, Wester PW, Vos JG, de Vries A & van Steeg H (2004) *p53* heterozygosity results in an increased 2-acetylaminofluorene-induced urinary bladder but not liver tumor response in DNA repair-deficient Xpa mice. Cancer Res, **64**: 5118–5126.

Hooker AM, Horne R, Morley AA & Sykes PJ (2002) Dose-dependent increase or decrease of somatic intrachromosomal recombination produced by etoposide. Mutat Res, **500**: 117–124.

Hooker AM, Bhat M, Day TK, Lane JM, Swinburne SJ, Morley AA & Sykes PJ (2004a) The linear no-threshold model does not hold for low dose ionizing radiation. Radiat Res, **162**: 447–452.

Hooker AM, Morley AA, Tilley WD & Sykes PJ (2004b) Cancer-associated genes can affect somatic intrachromosomal recombination early in carcinogenesis. Mutat Res, **550**: 1–10.

Hoorn AJW, Custer LL, Myhr BC, Brusick D, Gossen J & Vijg J (1993) Detection of chemical mutagens using Muta™Mouse: a transgenic mouse model. Mutagenesis, **8**: 7–10.

Horiguchi M, Masumura K, Ikehata H, Ono T, Kanke Y, Sofuni T & Nohmi T (1999) UVB-induced *gpt* mutations in the skin of *gpt* delta transgenic mice. Environ Mol Mutagen, **34**: 72–79.

Horiguchi M, Masumura K, Ikehata H, Ono T, Kanke Y & Nohmi T (2001) Molecular nature of ultraviolet B light-induced deletions in the murine epidermis. Cancer Res, **61**: 3913–3918.

Horsfall MJ & Glickman BW (1989) Mutational specificities of environmental carcinogens in the *lacI* gene of *Escherichia coli*. I. The direct-acting analogue *N*-nitroso-*N*-methyl-*N*-alpha-acetoxymethylamine. Carcinogenesis, **10**(5): 817–822.

Hoshi M, Morimura K, Wanibuchi H, Wei M, Okochi E, Ushijima T, Takaoka K & Fukushima S (2004) No-observed effect levels for carcinogenicity and for in vivo mutagenicity of a genotoxic carcinogen. Toxicol Sci, **81**(2): 273–279.

Hoyes KP, Wadeson PJ, Sharma HL, Hendry JH & Morris ID (1998) Mutation studies in *lacI* transgenic mice after exposure to radiation or cyclophosphamide. Mutagenesis, **13**: 607–612.

HSDB (2003a) 2-Acetylaminofluorene. CAS: 53-96-3. Bethesda, Maryland, United States Department of Health and Human Services, National Institutes of Health, National Library of Medicine, Hazardous Substances Data Bank (http://toxnet.nlm.nih.gov/cgi-bin/sis/htmlgen?HSDB).

HSDB (2003b) Quinoline. CAS: 91-22-5. Bethesda, Maryland, United States Department of Health and Human Services, National Institutes of Health, National Library of Medicine, Hazardous Substances Data Bank.

Hüttner E, Braun R & Schöneich J (1988) Mammalian spot test with the mouse for detection of transplacental genetic effects induced by 2-acetylaminofluorene and 4-acetylaminofluorene. In: Ashby J ed. Evaluation of short-term tests for carcinogens: report of the International Programme on Chemical Safety's collaborative study on in vivo assays. Cambridge, Cambridge University Press, pp 164–167.

IARC (1971) 4-Aminobiphenyl. In: Some inorganic substances, chlorinated hydrocarbons, aromatic amines, *N*-nitroso compounds, and natural products. Lyon, International Agency for Research on Cancer, pp 74–79 (IARC Monographs on the Evaluation of the Carcinogenic Risk of Chemicals to Man, Vol. 1).

IARC (1974a) Ethyl methanesulfonate. In: Some anti-thyroid and related substances, nitrofurans and industrial chemicals. Lyon, International Agency for Research on Cancer, pp 245–251 (IARC Monographs on the Evaluation of the Carcinogenic Risk of Chemicals to Man, Vol. 7).

IARC (1974b) Methyl methanesulfonate. In: Some anti-thyroid and related substances, nitrofurans and industrial chemicals. Lyon, International Agency for Research on Cancer, pp 253–260 (IARC Monographs on the Evaluation of the Carcinogenic Risk of Chemicals to Man, Vol. 7).

IARC (1974c) *N*-Methyl-*N*'-nitro-*N*-nitrosoguanidine. In: Some aromatic amines, hydrazine and related substances, *N*-nitroso compounds and miscellaneous alkylating agents. Lyon, International Agency for Research on Cancer, pp 183–195 (IARC Monographs on the Evaluation of the Carcinogenic Risk of Chemicals to Man, Vol. 4).

IARC (1974d) Urethane. In: Some anti-thyroid and related substances, nitrofurans and industrial chemicals. Lyon, International Agency for Research on Cancer, pp 111–140 (IARC Monographs on the Evaluation of the Carcinogenic Risk of Chemicals to Man, Vol. 7).

IARC (1976) Mitomycin C. In: Some naturally occurring substances. Lyon, International Agency for Research on Cancer, pp 171–179 (IARC Monographs on the Evaluation of the Carcinogenic Risk of Chemicals to Man, Vol. 10).

IARC (1977a) Asbestos. Lyon, International Agency for Research on Cancer, pp 1–106 (IARC Monographs on the Evaluation of the Carcinogenic Risk of Chemicals to Man, Vol. 14).

IARC (1977b) Phenobarbital and phenobarbital sodium. In: Some miscellaneous pharmaceutical substances. Lyon, International Agency for Research on Cancer, pp 157–181 (IARC Monographs on the Evaluation of the Carcinogenic Risk of Chemicals to Man, Vol. 13).

IARC (1978a) *N*-Nitrosodiethylamine. In: Some *N*-nitroso compounds. Lyon, International Agency for Research on Cancer, pp 89–106 (IARC Monographs on the Evaluation of the Carcinogenic Risk of Chemicals to Humans, Vol. 17).

IARC (1978b) *N*-Nitrosodimethylamine. In: Some *N*-nitroso compounds. Lyon, International Agency for Research on Cancer, pp 125–175 (IARC Monographs on the Evaluation of the Carcinogenic Risk of Chemicals to Humans, Vol. 17).

IARC (1978c) *N*-Nitroso-*N*-ethylurea. In: Some *N*-nitroso compounds. Lyon, International Agency for Research on Cancer, pp 191–215 (IARC Monographs on the Evaluation of the Carcinogenic Risk of Chemicals to Humans, Vol. 17).

IARC (1978d) *N*-Nitroso-*N*-methylurea. In: Some *N*-nitroso compounds. Lyon, International Agency for Research on Cancer, pp 227–255 (IARC Monographs on the Evaluation of the Carcinogenic Risk of Chemicals to Humans, Vol. 17).

IARC (1979) 1,2-Dichloroethane. In: Some halogenated hydrocarbons. Lyon, International Agency for Research on Cancer, pp 429–448 (IARC Monographs on the Evaluation of the Carcinogenic Risk of Chemicals to Humans, Vol. 20).

IARC (1980) Saccharin. In: Some non-nutritive sweetening agents. Lyon, International Agency for Research on Cancer, pp 132–170 (IARC Monographs on the Evaluation of the Carcinogenic Risk of Chemicals to Humans, Vol. 22).

IARC (1981a) Chlorambucil. In: Some antineoplastic and immunosuppressive agents. Lyon, International Agency for Research on Cancer, pp 115–136 (IARC Monographs on the Evaluation of the Carcinogenic Risk of Chemicals to Humans, Vol. 26).

IARC (1981b) Cyclophosphamide. In: Some antineoplastic and immunosuppressive agents. Lyon, International Agency for Research on Cancer, pp 165–202 (IARC Monographs on the Evaluation of the Carcinogenic Risk of Chemicals to Humans, Vol. 26).

IARC (1981c) Procarbazine hydrochloride. In: Some antineoplastic and immunosuppressive agents. Lyon, International Agency for Research on Cancer, pp 311–339 (IARC Monographs on the Evaluation of the Carcinogenic Risk of Chemicals to Humans, Vol. 26).

IARC (1982) Benzene. In: Some industrial chemicals and dyestuffs. Lyon, International Agency for Research on Cancer, pp 94–148 (IARC Monographs on the Evaluation of the Carcinogenic Risk of Chemicals to Humans, Vol. 29).

IARC (1983) Agaritine (*L*-glutamic acid-5-[2-(4-hydroxymethyl)-phenylhydrazide]). In: Some food additives, feed additives and naturally occurring substances. Lyon, International Agency for Research on Cancer, pp 63–69 (IARC Monographs on the Evaluation of the Carcinogenic Risk of Chemicals to Humans, Vol. 31).

IARC (1985) 4-(Methylnitrosamino)-1-(3-pyridyl)-1-butanone (NNK). In: Tobacco habits other than smoking. Lyon, International Agency for Research on Cancer, pp 209–223 (IARC Monographs on the Evaluation of the Carcinogenic Risk of Chemicals to Humans, Vol. 37).

IARC (1987a) 4,4'-Methylene bis(2-methylaniline) (Group 2B). In: Overall evaluations of carcinogenicity: an updating of IARC Monographs Volumes 1 to 42. Lyon, International Agency for Research on Cancer, pp 248–250 (IARC Monographs on the Evaluation of Carcinogenic Risks to Humans, Suppl. 7).

IARC (1987b) Asbestos (Group 1). In: Overall evaluations of carcinogenicity: an updating of IARC Monographs Volumes 1 to 42. Lyon, International Agency for Research on Cancer, pp 106–116 (IARC Monographs on the Evaluation of Carcinogenic Risks to Humans, Suppl. 7).

IARC (1987c) Asbestos. In: Genetic and related effects: an updating of selected IARC monographs from volumes 1 to 42. Lyon, International Agency for Research on Cancer, pp 77–80 (IARC Monographs on the Evaluation of Carcinogenic Risks to Humans, Suppl. 6).

IARC (1987d) Chlorambucil (Group 1). In: Overall evaluations of carcinogenicity: an updating of IARC Monographs Volumes 1 to 42. Lyon, International Agency for Research on Cancer, pp 144–145 (IARC Monographs on the Evaluation of Carcinogenic Risks to Humans, Suppl. 7).

IARC (1987e) Cyclophosphamide (Group 1). In: Overall evaluations of carcinogenicity: An updating of IARC Monographs Volumes 1 to 42. Lyon, International Agency for Research on Cancer, pp 182–184 (IARC Monographs on the Evaluation of Carcinogenic Risks to Humans, Suppl. 7).

IARC (1987f) *N*-Methyl-*N*'-nitro-*N*-nitrosoguanidine. In: Genetic and related effects: an updating of selected IARC monographs from volumes 1 to 42. Lyon, International Agency for Research on Cancer, 394–398 (IARC Monographs on the Evaluation of Carcinogenic Risks to Humans, Suppl. 6).

IARC (1987g) Phenobarbital (Group 2b). In: Overall evaluations of carcinogenicity: an updating of IARC Monographs Volumes 1 to 42. Lyon, International Agency for Research on Cancer, pp 313–316 (IARC Monographs on the Evaluation of Carcinogenic Risks to Humans, Suppl. 7).

IARC (1987h) Phenobarbital. In: Genetic and related effects: an updating of selected IARC monographs from volumes 1 to 42. Lyon, International Agency for Research on Cancer, pp 455–458 (IARC Monographs on the Evaluation of Carcinogenic Risks to Humans, Suppl. 6).

IARC (1987i) Procarbazine hydrochloride (Group 2A). In: Overall evaluations of carcinogenicity: an updating of IARC Monographs Volumes 1 to 42. Lyon, International Agency for Research on Cancer, pp 327–328 (IARC Monographs on the Evaluation of Carcinogenic Risks to Humans, Suppl. 7).

IARC (1987j) Saccharin (Group B). In: Overall evaluations of carcinogenicity: an updating of IARC Monographs Volumes 1 to 42. Lyon, International Agency for Research on Cancer, pp 334–339 (IARC Monographs on the Evaluation of Carcinogenic Risks to Humans, Suppl. 7).

IARC (1993a) Aflatoxins. In: Some naturally occurring substances: food items and constituents, heterocyclic aromatic amines and mycotoxins. Lyon, International Agency for Research on Cancer, pp 245–395 (IARC Monographs on the Evaluation of Carcinogenic Risks to Humans, Vol. 56).

IARC (1993b) IQ (2-Amino-3-methylimidazo[4,5-*f*]quinoline). In: Some naturally occurring substances: food items and constituents, heterocyclic aromatic amines and mycotoxins. Lyon, International Agency for Research on Cancer, pp 165–195 (IARC Monographs on the Evaluation of Carcinogenic Risks to Humans, Vol. 56).

IARC (1993c) MeIQ (2-Amino-3,4-dimethylinidazo[4,5-*f*]quinoline). In: Some naturally occurring substances: food items and constituents, heterocyclic aromatic amines and mycotoxins. Lyon, International Agency for Research on Cancer, pp 197–210 (IARC Monographs on the Evaluation of Carcinogenic Risks to Humans, Vol. 56).

IARC (1993d) PhIP (2-Amino-1-methyl-6-phenylimidazo[4,5-*b*]pyridine). In: Some naturally occurring substances: food items and constituents, heterocyclic aromatic amines and mycotoxins. Lyon, International Agency for Research on Cancer, pp 229–242 (IARC Monographs on the Evaluation of Carcinogenic Risks to Humans, Vol. 56).

IARC (1993e) MeIQx (2-Amino-3,8-dimethylimidazo[4,5-*f*]quinoxaline). In: Some naturally occurring substances: food items and constituents, heterocyclic aromatic amines and mycotoxins. Lyon, International Agency for Research on Cancer, pp 211–228 (IARC Monographs on the Evaluation of Carcinogenic Risks to Humans, Vol. 56).

IARC (1994a) Acrylamide. In: Some industrial chemicals. Lyon, International Agency for Research on Cancer, pp 389–433 (IARC Monographs on the Evaluation of Carcinogenic Risks to Humans, Vol. 60).

IARC (1994b) Ethylene oxide. In: Some industrial chemicals. Lyon, International Agency for Research on Cancer, pp 73–160 (IARC Monographs on the Evaluation of Carcinogenic Risks to Humans, Vol. 60).

IARC (1996) Tamoxifen. In: Some pharmaceutical drugs. Lyon, International Agency for Research on Cancer, pp 253–365 (IARC Monographs on the Evaluation of Carcinogenic Risks to Humans, Vol. 66).

IARC (1997) Polychlorinated dibenzo-*para*-dioxins. In: Polychlorinated dibenzo-*para*-dioxins and polychlorinated dibenzofurans. Lyon, International Agency for Research on Cancer, pp 194–219 (IARC Monographs on the Evaluation of Carcinogenic Risks to Humans, Vol. 69).

IARC (1999a) 1,3-Butadiene. In: Re-evaluation of some organic chemicals, hydrazine and hydrogen peroxide (part one). Lyon, International Agency for Research on Cancer, pp 109–225 (IARC Monographs on the Evaluation of Carcinogenic Risks to Humans, Vol. 71).

IARC (1999b) Methyl methanesulfonate. In: Re-evaluation of some organic chemicals, hydrazine and hydrogen peroxide (part three A). Lyon, International Agency for Research on Cancer, pp 1059–1078 (IARC Monographs on the Evaluation of Carcinogenic Risks to Humans, Vol. 71).

IARC (1999c) Saccharin and its salts. In: Some chemicals that cause tumours of the kidney or urinary bladder in rodents and some other substances. Lyon, International Agency for Research on Cancer, pp 517–624 (IARC Monographs on the Evaluation of Carcinogenic Risks to Humans, Vol. 73).

IARC (1999d) Ethylene dibromide. In: Re-evaluation of some organic chemicals, hydrazine and hydrogen peroxide (part two). Lyon, International Agency for Research on Cancer, p 641 (IARC Monographs on the Evaluation of Carcinogenic Risks to Humans, Vol. 71).

IARC (1999e) 1,2-Dibromo-3-chloropropane. In: Re-evaluation of some organic chemicals, hydrazine and hydrogen peroxide (part two). Lyon, International Agency for Research on Cancer, p 479 (IARC Monographs on the Evaluation of Carcinogenic Risks to Humans, Vol. 71).

IARC (1999f) 1,2-Dichloroethane. In: Re-evaluation of some organic chemicals, hydrazine and hydrogen peroxide (part two). Lyon, International Agency for Research on Cancer, pp 501–529 (IARC Monographs on the Evaluation of Carcinogenic Risks to Humans, Vol. 71).

IARC (2001) Chlordane and heptachlor. In: Some thyrotropic agents. Lyon, International Agency for Research on Cancer, pp 411–493 (IARC Monographs on the Evaluation of Carcinogenic Risks to Humans, Vol. 79).

Ikeda H, Shimizu T, Ukita T & Kumagai M (1995) A novel assay for illegitimate recombination in *Escherichia coli:* simulation of lambda bio transducing phage formation by ultra-violet light and its independence from RecA function. Adv Biophys, **31**:197–208.

IPCS (1993) Benzene. Geneva, World Health Organization, International Programme on Chemical Safety, pp 1–156 (Environmental Health Criteria 150).

IPCS (1995a) Methyl bromide. Geneva, World Health Organization, International Programme on Chemical Safety, pp 1–324 (Environmental Health Criteria 166).

IPCS (1995b) Tris(2,3-dibromopropyl) phosphate and bis(2,3-dibromopropyl) phosphate. Geneva, World Health Organization, International Programme on Chemical Safety, pp 1–129 (Environmental Health Criteria 173).

IPCS (1998a) Limonene. Geneva, World Health Organization, International Programme on Chemical Safety, pp 1–32 (Concise International Chemical Assessment Document No. 5).

IPCS (1998b) Selected non-heterocyclic polycyclic aromatic hydrocarbons. Geneva, World Health Organization, International Programme on Chemical Safety, pp 1–883 (Environmental Health Criteria 202).

IRIS (2002a) Bromomethane. CAS: 74-83-9. Bethesda, Maryland, United States Department of Health and Human Services, National Institutes of Health, National Library of Medicine, Integrated Risk Information System (http://toxnet.nlm.nih.gov/cgi-bin/sis/htmlgen?IRIS).

IRIS (2002b) *N*-Nitrosodi-*n*-propylamine. CAS: 621-64-7. Bethesda, Maryland, United States Department of Health and Human Services, National Institutes of Health, National Library of Medicine, Integrated Risk Information System (http://toxnet.nlm.nih.gov/cgi-bin/sis/htmlgen?IRIS).

IRIS (2002c) Quinoline. CAS: 91-22-5. Bethesda, Maryland, United States Department of Health and Human Services, National Institutes of Health, National Library of Medicine, Integrated Risk Information System (http://toxnet.nlm.nih.gov/cgi-bin/sis/htmlgen?IRIS).

Ishidate M & Odashima S (1977) Chromosome tests with 134 compounds on Chinese hamster cells in vitro — a screening for chemical carcinogens. Mutat Res, **48**(3–4): 337–354.

Itoh S, Miura M & Shimada H (1997) Germ cell mutagenesis in *lacZ* transgenic mice treated with methyl methanesulfonate. Mutat Res, **388**: 223–228.

Itoh S, Miura M, Itoh T, Miyauchi Y, Suga M, Takahashi Y, Kasahara Y, Yamamura E, Hirono H & Shimada H (1999) *N*-Nitrosodi-*n*-propylamine induces organ specific mutagenesis with specific expression times in *lacZ* transgenic mice. Mutat Res, **444**: 309–319.

Itoh T, Suzuki T, Nishikawa A, Furukawa F, Takahashi M, Xue W, Sofuni T & Hayashi M (2000) In vivo genotoxicity of 2-amino-3,8-dimethylimidazo[4,5-*f*]quinoline in *lacI* transgenic (Big Blue®) mice. Mutat Res, **468**: 19–25.

Itoh T, Kuwahara T, Suzuki T, Hayashi M & Ohnishi Y (2003) Regional mutagenicity of heterocyclic amines in the intestine: mutation analysis of the *cII* gene in lambda/*lacZ* transgenic mice. Mutat Res, **539**(1–2): 99–108.

Jakubczak JL, Merlino G, French JE, Muller WJ, Paul B, Adhya S & Garges S (1996) Analysis of genetic instability during mammary tumor progression using a novel selection-based assay for in vivo mutations in a bacteriophage λ transgene target. Proc Natl Acad Sci USA, **93**: 9073–9078.

JEMS/MMS (1996) Organ variation in the mutagenicity of ethylnitrosourea in Muta™Mouse: results of the collaborative study on the transgenic mutation assay by JEMS/MMS. Environ Mol Mutagen, **28**: 363–375.

Jenssen D & Ramel C (1976) Dose response at low doses of X-irradiation and MMS on the induction of micronuclei in mouse erythroblasts. Mutat Res, **41**: 311–319.

Jenssen D & Ramel C (1980) The micronucleus test as part of a short-term mutagenicity test program for the prediction of carcinogenicity evaluated by 143 agents tested. Mutat Res, **75**: 191–202.

Kalnins A, Otto K, Rüther U & Müller-Hill B (1983) Sequence of the *lacZ* gene of *Escherichia coli*. EMBO J, **2**: 593–597.

Katoh M, Horiya N & Valdivia RPA (1997) Mutations induced in male germ cells after treatment of transgenic mice with ethylnitrosourea. Mutat Res, **388**: 229–237.

Kliesch U, Danford N & Adler I-D (1981) Micronucleus test and bone-marrow chromosome analysis: a comparison of 2 methods in vivo for evaluating chemically induced chromosomal alterations. Mutat Res, **80**(2): 321–332.

Klungland A, Rosewell I, Hollenbach S, Larsen E, Daly G, Epe B, Seeberg E, Lindahl T & Barnes DE (1999) Accumulation of premutagenic DNA lesions in mice defective in removal of oxidative base damage. Proc Natl Acad Sci USA, **96**: 13300–13305.

Kohara A, Suzuki T, Honma M, Hirano N, Ohsawa K, Ohwada T & Hayashi M (2001) Mutation spectrum of *o*-aminoazotoluene in the *cII* gene of lambda/*lacZ* transgenic mice (Muta™Mouse). Mutat Res, **491**(1–2): 211–220.

Kohara A, Suzuki T, Honma M, Ohwada T & Hayashi M (2002a) Mutagenicity of aristolochic acid in the *λ/lacZ* transgenic mouse (Muta™Mouse). Mutat Res, **515**(1–2): 63–72.

Kohara A, Suzuki T, Honma M, Oomori T, Ohwada T & Hayashi M (2002b) Dinitropyrenes induce gene mutations in multiple organs of the *λ/lacZ* transgenic mouse (Muta™Mouse). Mutat Res, **515**(1–2): 73–83.

Kohler SW, Provost GS, Kretz PL, Fieck A, Sorge JA & Short JM (1990) The use of transgenic mice for short-term, in vivo mutagenicity testing. Genet Anal Tech Appl, **7**: 212–218.

Kohler SW, Provost GS, Fieck A, Kretz PL, Bullock WO, Putman DL, Sorge JA & Short JM (1991a) Analysis of spontaneous and induced mutations in transgenic mice using a lambda ZAP/*lacI* shuttle vector. Environ Mol Mutagen, **18**: 316–321.

Kohler SW, Provost GS, Fieck A, Kretz PL, Bullock WO, Sorge JA, Putman DL & Short JM (1991b) Spectra of spontaneous and mutagen-induced mutations in the *lacI* gene in transgenic mice. Proc Natl Acad Sci USA, **88**: 7958–7962.

Kosinska W, von Pressentin MdM & Guttenplan JB (1999) Mutagenesis induced by benzo[*a*]pyrene in *lacZ* mouse mammary and oral tissues: comparisons with mutagenesis in other organs and relationships to previous carcinogenicity assays. Carcinogenesis, **20**(6): 1103–1106.

Krebs O & Favor J (1997) Somatic and germ cell mutagenesis in lambda *lacZ* transgenic mice treated with acrylamide or ethylnitrosourea. Mutat Res, **388**: 239–248.

Leach EG, Narayanan L, Havre PA, Gunther EJ, Yeasky TM & Glazer PM (1996a) Tissue specificity of spontaneous point mutations in lambda *supF* transgenic mice. Environ Mol Mutagen, **28**: 459–464.

Leach EG, Gunther EJ, Yeasky TM, Gibson LH, Yang-Feng TL & Glazer PM (1996b) Frequent spontaneous deletions at a shuttle vector locus in transgenic mice. Mutagenesis, **11**(1): 49–56.

Lefevre PA, Tinwell H & Ashby J (1997) Mutagenicity of the potent rat hepatocarcinogen 6BT to the liver of transgenic (*lacI*) rats: consideration of a reduced mutation assay protocol. Mutagenesis, **12**: 45–47.

Limoli CL, Giedzinski E, Bonner WM & Cleaver JE (2002) UV-induced replication arrest in the xeroderma pigmentosum variant leads to DNA double-strand breaks, gamma-H2AX formation, and Mre11 relocalization. Proc Natl Acad Sci USA, **99**(1): 233–238.

Lonardo EC, Perdue PA & Freeman JJ (1996) The detection of gene mutation in transgenic mice (Big Blue®) following administration of a known mutagen 7,12-dimethylbenz[*a*]anthracene — (DMBA). Abstr Annu Meet Environ Mutagen Soc, **1996**: 42.

Loprieno N, Boncristiani G & Loprieno G (1991) An experimental approach to identifying the genotoxic risk from cooked meat mutagens. Food Chem Toxicol, **29**(6): 377–386.

Loveday KS, Anderson BE, Resnick MA & Zeiger E (1990) Chromosome aberration and sister chromatid exchange tests in Chinese hamster ovary cells in vitro. V: Results with 46 chemicals. Environ Mol Mutagen, **16**: 272–303.

Lynch AM, Gooderham NJ, Davies DS & Boobis AR (1998) Genetic analysis of PHIP intestinal mutations in Muta™Mouse. Mutagenesis, **13**(6): 601–605.

Machemer L & Lorke D (1978) Mutagenicity studies with praziquantel, a new anthelmintic drug, in mammalian systems. Arch Toxicol, **39**: 187–197.

MAK (1976) β-Propiolacton. In: Gesundheitsschädliche Arbeitsstoffe: Toxikologisch-arbeitsmedizinische Begründung von MAK-Werten. Vol. 9. Deutsche Forschungsgemeinschaft (DFG). Weinheim, Wiley-VCH, pp 1–8.

MAK (1991) *N*-Nitrosamines. In: Occupational toxicants — critical data evaluation for MAK values and classification of carcinogens. Vol. 1. Deutsche Forschungsgemeinschaft (DFG). Weinheim, Wiley-VCH, pp 261–289.

MAK (1994) Toluene-2,4-diamine. In: Occupational toxicants — critical data evaluation for MAK values and classification of carcinogens. Vol. 6. Deutsche Forschungsgemeinschaft (DFG). Weinheim, Wiley-VCH, pp 339–352.

MAK (1998) Trichloroethylene. In: Occupational toxicants — critical data evaluation for MAK values and classification of carcinogens. Vol. 10. Deutsche Forschungsgemeinschaft (DFG). Weinheim, Wiley-VCH, pp 221–236.

MAK (2000) Chloroform. In: Occupational toxicants — critical data evaluation for MAK values and classification of carcinogens. Vol. 14. Deutsche Forschungsgemeinschaft (DFG). Weinheim, Wiley-VCH, pp 19–58.

MAK (2002a) Di(2-ethylhexyl)phthalat (DEHP). In: Gesundheitsschädliche Arbeitsstoffe: Toxikologisch-arbeitsmedizinische Begründung von MAK-Werten. Vol. 4. Deutsche Forschungsgemeinschaft (DFG). Weinheim, Wiley-VCH, pp 1–81.

MAK (2002b) Carbon tetrachloride. In: Occupational toxicants — critical data evaluation for MAK values and classification of carcinogens. Vol. 18. Deutsche Forschungsgemeinschaft (DFG). Weinheim, Wiley-VCH, pp 82–106.

Malling HV & Delongchamp RR (2001) Direct separation of in vivo and in vitro *am3* revertants in transgenic mice carrying the ΦX174 *am3*, *cs70* vector. Environ Mol Mutagen, **37**: 345–355.

Malling HV, Delongchamp RR & Valentine CR (2003) Three origins of ΦX174 *am3* revertants in transgenic cell culture. Environ Mol Mutagen, **42**: 258–273.

Manjanatha MG, Shelton SD, Aidoo A, Lyn-Cook LE & Casciano DA (1998) Comparison of in vivo mutagenesis in the endogenous *Hprt* gene and the *lacI* transgene of Big Blue® rats treated with 7,12-dimethylbenz[*a*]anthracene. Mutat Res, **401**: 165–178.

Manjanatha MG, Shelton SD, Culp SJ, Blankenship LR & Casciano DA (2000) DNA adduct formation and molecular analysis of in vivo *lacI* mutations in the mammary tissue of Big Blue® rats treated with 7,12-dimethylbenz[*a*]anthracene. Carcinogenesis, **21**(2): 265–273.

Marshall E (2001) Genome sequencing. Celera assembles mouse genome; public labs plan new strategy. Science, **292**: 822–823.

Masumura K, Matsui M, Katoh M, Horiya N, Ueda O, Tanabe H, Yamada M, Suzuki H, Sofuni T & Nohmi T (1999a) Spectra of *gpt* mutations in ethylnitrosourea-treated and untreated transgenic mice. Environ Mol Mutagen, **34**: 1–8.

Masumura K, Matsui K, Yamada M, Horiguchi M, Ishida K, Watanabe M, Ueda O, Suzuki H, Kanke Y, Tindall KR, Wakabayashi K, Sofuni T & Nohmi T (1999b) Mutagenicity of 2-amino-1-methyl-6-phenylimidazo[4,5-*b*]pyridine (PhIP) in the new *gpt* δ-transgenic mouse. Cancer Lett, **143**: 241–244.

Masumura K, Matsui K, Yamada M, Horiguchi M, Ishida K, Watanabe M, Wakabayashi K & Nohmi T (2000) Characterization of mutations induced by 2-amino-1-methyl-6-phenyl-imidazo[4,5-*b*]pyridine in the colon of *gpt* delta transgenic mouse: novel G:C deletions beside runs of identical bases. Carcinogenesis, **21**(11): 2049–2056.

Masumura K, Kuniya K, Kurobe T, Fukuoka M, Yatagai F & Nohmi T (2002) Heavy-ion-induced mutations in the *gpt* delta transgenic mouse: Comparison of mutation spectra induced by heavy-ion, X-ray, and γ-ray radiation. Environ Mol Mutagen, **40**(3): 207–215.

Matsuoka A, Hayashi M & Ishidate M (1979) Chromosomal aberration tests on 29 chemicals combined with S9 mix in vitro. Mutat Res, **66**(3): 277–290.

Matsuoka F, Nagawa F, Okazaki K, Kingsbury L, Yoshida K, Muller U, Larue DT, Winter JA & Sakano H (1991) Detection of somatic DNA recombination in the transgenic mouse brain. Science, **254**: 81–86.

Mavournin KH, Blakey DH, Cimino MC, Salamone MF & Heddle JA (1990) The in vivo micronucleus assay in mammalian bone marrow and peripheral blood. A report of the U.S. Environmental Protection Agency GENE-TOX Program. Mutat Res, **239**: 29–80.

Mei N, Heflich RH, Chou MW & Chen T (2004) Mutations induced by the carcinogenic pyrrolizidine alkaloid riddelline in the liver *cII* gene of transgenic Big Blue® rats. Chem Res Toxicol, **17**(6): 814–818.

Meyne J, Allison DC, Bose K, Jordan SW, Ridolpho PF & Smith J (1985) Hepatotoxic doses of dioxin do not damage mouse bone marrow chromosomes. Mutat Res, **157**(1): 63–69.

Micillino JC, Coulais C, Binet S, Bottin M-C, Keith G, Moulin D & Rihn BH (2002) Lack of genotoxicity of bitumen fumes in transgenic mouse lung. Toxicology, **170**(1–2): 11–20.

Mientjes EJ, Steenwinkel MJST, van Delft JHM, Lohman PHM & Baan RA (1996) Comparison of the X-gal- and P-gal-based systems for screening of mutant *λlacZ* phages originating from the transgenic mouse strain 40.6. Mutat Res, **360**: 101–106.

Mientjes EJ, Luiten-Schuite A, van der Wolf E, Borsboom Y, Bergmanns A, Berends F, Lohman PHM, Baan RA & van Delft JHM (1998) DNA adducts, mutant frequencies, and mutation spectra in various organs of *λlacZ* mice exposed to ethylating agents. Environ Mol Mutagen, **31**: 18–31.

Minowa O, Arai T, Hirano M, Monden Y, Nakai S, Fukuda M, Itoh M, Takano H, Hippou Y, Aburatani H, Masumura K, Nohmi T, Nishimura S & Noda T (2000) *Mmh/Ogg1* gene inactivation results in accumulation of 8-hydroxyguanine in mice. Proc Natl Acad Sci USA, **97**: 4156–4161.

Mirsalis JC, Provost GS, Matthews CD, Hamner RT, Schindler JE, O'Loughlin KG, MacGregor JT & Short JM (1993) Induction of hepatic mutations in *lacI* transgenic mice. Mutagenesis, **8**: 265–271.

Mirsalis J, Monforte J & Winegar R (1995) Transgenic animal models for detection of in vivo mutations. Annu Rev Pharm Toxicol, **35**: 145–164.

Mittelstaedt RA, Manjanatha MG, Shelton SD, Lyn-Cook LE, Chen JB, Aidoo A, Casciano DA & Heflich RH (1998) Comparison of the types of mutations induced by 7,12-dimethylbenz[*a*]anthracene in the *lacI* and *Hprt* genes of Big Blue® rats. Environ Mol Mutagen, **31**: 149–156.

Mittelstaedt RA, Mei N, Webb PJ, Shaddock JG, Dobrovolsky VN, McGarrity LJ, Morris SM, Chen T, Beland FA, Greenlees KJ & Heflich RH (2004) Genotoxicity of malachite green and leucomalachite green in female Big Blue® B6C3F1 mice. Mutat Res, **561**(1–2): 127–138.

Miyata Y, Saeki K, Kawazoe Y, Hayashi M, Sofuni T & Suzuki T (1998) Antimutagenic structural modification of quinoline assessed by an in vivo mutagenesis assay using *lacZ*¯ transgenic mice. Mutat Res, **414**: 165–169.

Møller P, Wallin H, Vogel U, Autrup H, Risom L, Hald MT, Daneshvar B, Dragsted LO, Poulsen HE & Loft S (2002) Mutagenicity of 2-amino-3-methylimidazol[4,5-*f*]quinoline in colon and liver of Big Blue® rats: role of DNA adducts, strand breaks, DNA repair and oxidative stress. Carcinogenesis, **23**(8): 1379–1385.

Monroe JJ, Kort KL, Miller JE, Marino DR & Skopek TR (1998) A comparative study of in vivo mutation assays: analysis of *Hprt*, *lacI*, *cII/cI* as mutational targets for *N*-nitroso-*N*-methylurea and benzo[*a*]pyrene in Big Blue® mice. Mutat Res, **421**: 121–136.

Morita T, Asano N, Awogi T, Sasaki Y, Sato S, Shimada H, Sutou S, Suzuki T, Wakata A, Sofuni T & Hayashi M (1997a) Evaluation of the rodent micronucleus assay in the screening of IARC carcinogens (groups 1, 2A and 2B): The summary report of the 6th collaborative study by CSGMT/JEMS MMS. Collaborative Study of the Micronucleus Group Test. Mammalian Mutagenicity Study Group. Mutat Res, **389**(1): 3–122.

Morita T, Asano N, Awogi T, Sasaki Y, Sato S, Shimada H, Sutou S, Suzuki T, Wakata A, Sofuni T & Hayashi M (1997b) Erratum to "Evaluation of the rodent micronucleus assay in the screening of IARC carcinogens (groups 1, 2A and 2B): The summary report of the 6th collaborative study by CSGMT/JEMS MMS." Mutat Res, **391**: 259–267.

Morris HP, Velat CA, Wagner BP, Dahlgard M & Ray FE (1960) Studies of carcinogenicity in the rate of derivates of aromatic amines related to *N*-2-fluorenylacetamide. J Natl Cancer Inst, **24**: 149–180.

Müller AK, Farombi EO, Moller P, Autrup HN, Vogel U, Wallin H, Dragsted LO, Loft S & Binderup ML (2004) DNA damage in lung after oral exposure to diesel exhaust particles in Big Blue® rats. Mutat Res, **550**(1–2): 123–132.

Mullin AH, Rando R, Esmundo F & Mullin DA (1995) Inhalation of benzene leads to an increase in the mutant frequencies of a *lacI* transgene in lung and spleen tissues of mice. Mutat Res, **327**: 121–129.

Mullin AH, Nataraj D, Ren J-J & Mullin DA (1998) Inhaled benzene increases the frequency and length of *lacI* deletion mutations in lung tissues of mice. Carcinogenesis, **19**(10): 1723–1733.

Myhr B (1991) Validation studies with Muta Mouse: a transgenic mouse model for detecting mutations in vivo. Environ Mol Mutagen, **18**(4): 308–315.

Nagao M (1999) A new approach to risk estimation of food-borne carcinogens heterocyclic amines — based on molecular information. Mutat Res, **431**(1): 3–12.

Nagao M, Fujita H, Ochiai M, Wakabayashi K, Sofuni T, Matsushima T, Sugimura T & Ushijima T (1998) No direct correlation between mutant frequencies and cancer incidence induced by MeIQ in various organs of Big Blue® mice. Mutat Res, **400**: 251–257.

Nagao M, Ochiai M, Okochi E, Ushijima T & Sugimura T (2001) *LacI* transgenic animal study: relationships among DNA-adduct levels, mutant frequencies and cancer incidences. Mutat Res, **477**: 119–124.

Nakajima M, Kikuchi M, Saeki K, Miyata Y, Terada M, Kishida F, Yamamoto R, Furihata C & Dean SW (1999) Mutagenicity of 4-nitroquinoline 1-oxide in the Muta™Mouse. Mutat Res, **444**: 321–336.

Neuhäuser A (1977) Die Wirksamkeit von Natulan im Fellflecken-Test mit der Maus. GSF-Ber B, **798**: 42–44.

Neuhäuser-Klaus A & Chauhan P (1987) Studies on somatic mutation induction in the mouse with isoniazid and hydrazine. Mutat Res, **191**: 111–116.

Neuhäuser-Klaus A & Schmahl W (1989) Mutagenic and teratogenic effects of acrylamide in the mammalian spot test. Mutat Res, **226**(3): 157–162.

Nishino H, Buettner VL, Haavik J, Schaid DJ & Sommer SS (1996) Spontaneous mutation in Big Blue® transgenic mice: analysis of age, gender and tissue type. Environ Mol Mutagen, **28**(4): 299–312.

Nohmi T & Masumura K (2004) *gpt* delta transgenic mouse: a novel approach for molecular dissection of deletion mutations in vivo. Adv Biophys, **38**: 97–121.

Nohmi T & Masumura K (2005) Molecular nature of intrachromosomal deletions and base substitutions induced by environmental mutagens. Environ Mol Mutagen, **45**: 150–161.

Nohmi T, Katoh M, Suzuki H, Matsui M, Yamada M, Watanabe M, Suzuki M, Horiya N, Ueda O, Shibuya T, Ikeda H & Sofuni T (1996) A new transgenic mouse mutagenesis test system using Spi^- and 6-thioguanine selections. Environ Mol Mutagen, **28**: 465–470.

Nohmi T, Suzuki M, Masumura K, Yamada M, Matsui K, Ueda O, Suzuki H, Katoh M, Ikeda H & Sofuni T (1999) Spi^- selection: An efficient method to detect γ-ray-induced deletions in transgenic mice. Environ Mol Mutagen, **34**: 9–15.

Nohmi T, Suzuki T & Masumura K (2000) Recent advances in the protocols of transgenic mouse mutation assays. Mutat Res, **455**(1–2): 191–215.

NTP (2003a) 2-Acetylaminofluorene. CAS: 53-96-3. Research Triangle Park, North Carolina, United States Department of Health and Human Services, National Toxicology Program (Factsheet; http://ntp-server.niehs.nih.gov/index.cfm?objectid=6DDF4AD9-F1F6-975E-7A0747B910250667; last updated 4 September 2003; accessed 11 November 2004).

NTP (2003b) 2,6-Toluenediamine dihydrochloride (2,6-diaminotoluene dihydrochloride). CAS 15481-70-6. Research Triangle Park, North Carolina, United States Department of Health and Human Services, National Toxicology Program (Factsheet: http://ntp.niehs.nih.gov/index.cfm?objectid=0718C4BA-E9A2-42AB-A9634AF5BFA7D317; last updated 4 September 2003; accessed 11 November 2004).

Ochiai M, Ishida K, Ushijima T, Suzuki T, Sofuni T, Sugimura T & Nagao M (1998) DNA adduct level induced by 2-amino-3,4-dimethylimidazo[4,5-*f*]-quinoline in Big Blue® mice does not correlate with mutagenicity. Mutagenesis, **13**(4): 381–384.

OECD (1971) OECD guideline for the testing of chemicals. OECD 486: Unscheduled DNA synthesis (UDS) test with mammalian liver cells in vivo. Paris, Organisation for Economic Co-operation and Development, pp 1–8.

OECD (1981) OECD guideline for the testing of chemicals. OECD 451: Carcinogenicity studies. Paris, Organisation for Economic Co-operation and Development, pp 1–17.

OECD (1984a) OECD guideline for the testing of chemicals. OECD 477: Genetic toxicology: Sex-linked recessive lethal test in *Drosophila melanogaster.* Paris, Organisation for Economic Co-operation and Development, pp 1–6.

OECD (1984b) OECD guideline for the testing of chemicals. OECD 478: Genetic toxicology: Rodent dominant lethal test. Paris, Organisation for Economic Co-operation and Development, pp 1–6.

OECD (1986a) OECD guideline for the testing of chemicals. OECD 484: Genetic toxicology: mouse spot test. Paris, Organisation for Economic Co-operation and Development, pp 1–4.

OECD (1986b) OECD guideline for the testing of chemicals. OECD 485: Genetic toxicology: Mouse heritable translocation assay. Paris, Organisation for Economic Co-operation and Development, pp 1–6.

OECD (1997a) OECD guideline for the testing of chemicals. OECD 474: Mammalian erythrocyte micronucleus test. Paris, Organisation for Economic Co-operation and Development, pp 1–10.

OECD (1997b) OECD guideline for the testing of chemicals. OECD 475: Mammalian bone marrow chromosome aberration test. Paris, Organisation for Economic Co-operation and Development, pp 1–8.

OECD (1997c) OECD guideline for the testing of chemicals. OECD 483: Mammalian spermatogonial chromosome aberration test. Paris, Organisation for Economic Co-operation and Development, pp 1–8.

Ohsawa K, Hirano N, Sugiura M, Nakagawa S & Kimura M (2000) Genotoxicity of *o*-aminoazotoluene (AAT) determined by the Ames test, the in vitro chromosomal aberration test, and the transgenic mouse gene mutation assay. Mutat Res, **471**: 113–126.

Okada N, Honda A, Kawabata M & Yajima N (1997) Sodium phenobarbital-enhanced mutation frequency in the liver DNA of *lacZ* transgenic mice treated with diethylnitrosamine. Mutagenesis, **12**: 179–184.

Okochi E, Watanabe N, Shimada Y, Takahashi S, Wakazono K, Shirai T, Sugimura T, Nagao M & Ushijima T (1999) Preferential induction of guanine deletion at 5'-GGGA-3' in rat mammary glands by 2-amino-1-methyl-6-phenylimidazo[4,5-*b*]pyridine. Carcinogenesis, **20**(10): 1933–1938.

Okonogi H, Stuart GR, Okochi E, Ushijima T, Sugimura T, Glickman BW & Nagao M (1997) Effects of gender and species on spectra of mutation induced by 2-amino-1-methyl-6-phenylimidazo[4,5-*b*]pyridine in the *lacI* transgene. Mutat Res, **395**: 93–99.

Paashuis-Lew YR & Heddle JA (1998) Spontaneous mutation during fetal development and post-natal growth. Mutagenesis, **13**(6): 613–617.

Paashuis-Lew Y, Zhang XB & Heddle JA (1997) On the origin of spontaneous somatic mutations and sectored plaques detected in transgenic mice. Mutat Res, **373**(2): 277–284.

Piegorsch WW, Lockhart AC, Carr GJ, Margolin BH, Brooks T, Douglas GR, Liegibel UM, Suzuki T, Thybaud V, van Delft JHM & Gorelick NJ (1997) Sources of variability in data from a positive selection *lacZ* transgenic mouse mutation assay: an interlaboratory study. Mutat Res, **388**: 249–289.

Pletsa V, Valavanis C, van Delft JHM, Steenwinkel M-JST & Kyrtopoulos SA (1997) DNA damage and mutagenesis induced by procarbazine in *λlacZ* transgenic mice: evidence that bone marrow mutations do not arise primarily through miscoding by *O*^{6}-methylguanine. Carcinogenesis, **18**: 2191–2196.

Pletsa V, Steenwinkel MJ, van Delft JH, Baan RA & Kyrtopoulos SA (1999) Methyl bromide causes DNA methylation in rats and mice but fails to induce somatic mutations in lambda *lacZ* transgenic mice. Cancer Lett, **135**(1): 21–27.

Provost GS & Short JM (1994) Characterization of mutations induced by ethylnitrosourea in seminiferous tubule germ cells of transgenic B6C3F1 mice. Proc Natl Acad Sci USA, **91**: 6564–6568.

Provost GS, Kretz P, Hamner RT, Matthews CD, Rogers BJ, Lundberg KS, Dycaico MJ & Short JM (1993) Transgenic systems for in vivo mutation analysis. Mutat Res, **288**: 133–149.

Provost GS, Mirsalis JC, Rogers BJ & Short JM (1996) Mutagenic response to benzene and tris(2,3-dibromopropyl)-phosphate in the lambda *lacI* transgenic mouse mutation assay: a standardized approach to in vivo mutation analysis. Environ Mol Mutagen, **28**: 342–347.

Putman DL, Penn Ritter A, Carr GJ & Young RR (1997) Evaluation of spontaneous and chemical-induced *lacI* mutations in germ cells from λ/*lacI* transgenic mice. Mutat Res, **388**: 137–143.

RamaKrishna NVS, Devanesan PD, Rogan EG, Cavalieri EL, Jeong H, Jankowiak R & Small GJ (1992) Mechanism of metabolic activation of the potent carcinogen 7,12-dimethylbenz[a]anthracene. Chem Res Toxicol, **5**: 220–226.

Recio L, Bond JA, Pluta LJ & Sisk SC (1993) Use of transgenic mice for assessing the mutagenicity of 1,3-butadiene in vivo. In: Sorsa M, Peltonen K, Vainio H & Hemminki K eds. Butadiene and styrene: assessment of health hazards. Lyon, International Agency for Research on Cancer, pp 235–243 (IARC Scientific Publications No. 127).

Recio L, Meyer KG, Pluta LJ, Moss OR & Saranko CJ (1996) Assessment of 1,3-butadiene mutagenicity in the bone marrow of B6C3F1 *lacI* transgenic mice (Big Blue®): a review of mutational spectrum and *lacI* mutant frequency after a 5-day 625 ppm 1,3-butadiene exposure. Environ Mol Mutagen, **28**: 424–429.

Recio L, Donner M, Abernethy D, Pluta L, Steen AM, Wong BA, James A & Preston RJ (2004) In vivo mutagenicity and mutation spectrum in the bone marrow and testes of B6C3F1 *lacI* transgenic mice following inhalation exposure to ethylene oxide. Mutagenesis, **19**(3): 215–222.

Renault D, Brault D & Thybaud V (1997) Effect of ethylnitrosourea and methyl methanesulfonate on mutation frequency in Muta™Mouse germ cells (seminiferous tubule cells and epididymis spermatozoa). Mutat Res, **388**: 145–153.

Rihn B, Coulais C, Kauffer E, Bottin M-C, Martin P, Yvon F, Vigneron JC, Binet S, Monhoven N, Steiblen G & Keith G (2000) Inhaled crocidolite mutagenicity in lung DNA. Environ Health Perspect, **108**: 341–346.

RIVM (2000) Mutagenicity of chemicals in genetically modified animals. Bilthoven, National Institute of Public Health and the Environment (RIVM), pp 1–47 (RIVM Report No. 650210 002; TNO Report No. V99.1097).

Robbiano L, Mereto E, Morando AM, Pastore P & Brambilla G (1998) Increased frequency of micronucleated kidney cells in rats exposed to halogenated anaesthetics. Mutat Res, **413**: 1–6.

Rogers BJ, Provost GS, Young RR, Putman DL & Short JM (1995) Intralaboratory optimization and standardization of mutant screening conditions used for a lambda/*lacI* transgenic mouse mutagenesis assay (I). Mutat Res, **327**: 57–66.

Ross JA & Leavitt SA (1998) Induction of mutations by 2-acetylaminofluorene in *lacI* transgenic B6C3F1 mouse liver. Mutagenesis, **13**: 173–179.

Russell LB, Selby PB, von Halle E, Sheridan W & Valcovic L (1981) Use of the mouse spot test in chemical mutagenesis: interpretation of past data and recommendations for future work. Mutat Res, **86**: 355–379.

Sanger F, Coulson AR, Friedman T, Air GM, Barrell BG, Brown NL, Fiddes JC, Hutchison CA III, Slocombe PM & Smith M (1978) The nucleotide sequence of bacteriophage ΦX174. J Mol Biol, **125**: 225–246.

Sasaki YF, Imanishi H, Watanabe M, Sekiguchi A, Moriya M, Shirasu Y & Tutikawa K (1986) Mutagenicity of 1,2-dibromo-3-chloropropane (DBCP) in the mouse spot test. Mutat Res, **174**(2): 145–147.

Schaaper RM, Danforth BN & Glickman BW (1986) Mechanisms of spontaneous mutagenesis: an analysis of the spectrum of spontaneous mutation in the *Escherichia coli lacI* gene. J Mol Biol, **189**(2): 273–284.

Schaaper RM, Koffel-Schwartz N & Fuchs RP (1990) *N*-Acetoxy-*N*-acetyl-2-aminofluorene-induced mutagenesis in the *lacI* gene of *Escherichia coli*. Carcinogenesis, **11**(7): 1087–1095.

Schinz HR, Fritz-Niggli H, Campbell TW & Schmid H (1955) Krebsbildung durch Aminofluorene und verwandte Körper. Oncologia, **8**: 233–245.

Schmezer P & Eckert C (1999) Induction of mutations in transgenic animal models: Big Blue® and Muta™Mouse. In: McGregor DB, Rice JM & Venitt S eds. The use of short- and medium-term tests for carcinogens and data on genetic effects in carcinogenic hazard evaluation. Lyon, International Agency for Research on Cancer, pp 367–394 (IARC Scientific Publications No. 146).

Schmezer P, Eckert C, Liegibel U, Zelezny O & Klein R (1998a) Mutagenic activity of carcinogens detected in transgenic rodent mutagenicity assays at dose levels used in chronic rodent cancer bioassays. Mutat Res, **405**: 193–198.

Schmezer P, Eckert C, Liegibel U, Klein R & Bartsch H (1998b) Use of transgenic mutational test systems in risk assessment of carcinogens. Arch Toxicol Suppl, **20**: 321–330.

Shane BS, Lockhart AM, Winston GW & Tindall KR (1997) Mutant frequency of *lacI* in transgenic mice following benzo[*a*]pyrene treatment and partial hepatectomy. Mutat Res, **377**: 1–11.

Shane BS, de Boer J, Watson DE, Haseman JK, Glickman BW & Tidall KR (2000a) *LacI* mutation spectra following benzo[*a*]pyrene treatment of Big Blue® mice. Carcinogenesis, **21**: 715–725.

Shane BS, Smith-Dunn DL, de Boer JG, Glickman BW & Cunningham ML (2000b) Mutant frequencies and mutation spectra of dimethylnitrosamine (DMN) at the *lacI* and *cII* loci in the livers of Big Blue® transgenic mice. Mutat Res, **452**: 197–210.

Shane BS, Smith-Dunn DL, deBoer JG, Glickman BW & Cunningham ML (2000c) Subchronic administration of phenobarbital alters the mutation spectrum of *lacI* in the livers of Big Blue® transgenic mice. Mutat Res, **448**: 69–80.

Shaver-Walker PM, Urlando C, Tao KS, Zhang XB & Heddle JA (1995) Enhanced somatic mutation rates induced in stem cells of mice by low chronic exposure to ethylnitrosourea. Proc Natl Acad Sci USA, **92**: 11470–11474.

Shelby MD & Witt KL (1995) Comparison of results from mouse bone marrow chromosome aberration and micronucleus tests. Environ Mol Mutagen, **25**: 302–313.

Shelby MD, Erexson GL, Hook GJ & Tice RR (1993) Evaluation of a three-exposure mouse bone marrow micronucleus protocol: results with 49 chemicals. Environ Mol Mutagen, **21**: 160–179.

Shelton SD, Cherry V & Manjanatha MG (2000) Mutant frequency and molecular analysis of in vivo *lacI* mutations in the bone marrow of Big Blue® rats treated with 7,12-dimethylbenz[*a*]anthracene. Environ Mol Mutagen, **36**: 235–242.

Shephard SE, Sengstag C, Lutz WK & Schlatter C (1993) Mutations in liver DNA of *lacI* transgenic mice (Big Blue®) following subchronic exposure to 2-acetylaminofluorene. Mutat Res, **302**: 91–96.

Shephard SE, Gunz D & Schlatter C (1995) Genotoxicity of agaritine in the *lacI* transgenic mouse mutation assay: evaluation of the health risk of mushroom consumption. Food Chem Toxicol, **33**: 257–264.

Shibata A, Kamada N, Masumura K, Nohmi T, Kobayashi S, Teraoka H, Nakagama H, Sugimura T, Suzuki H & Masutani M (2005) *Parp-1* deficiency causes an increase of deletion mutations and insertions/rearrangements *in vivo* after treatment with an alkylating agent. Oncogene, **24**(8): 1328–1337.

Shibuya T & Morimoto K (1993) A review of the genotoxicity of 1-ethyl-1-nitrosourea. Mutat Res, **297**(1): 3–38.

Shimada H, Suzuki H, Itoh S, Hattori C, Matsuura Y, Tada S & Watanabe C (1992) The micronucleus test of benzo[*a*]pyrene with mouse and rat peripheral blood reticulocytes. Mutat Res, **278**: 165–168.

Singh VK, Ganesh L, Cunningham ML & Shane BS (2001) Comparison of the mutant frequencies and mutation spectra of three non-genotoxic carcinogens, oxazepam, phenobarbital, and Weyth 14,643, at the lambda*cII* locus in Big Blue® transgenic mice. Biochem Pharmacol, **62**(6): 685–692.

Sisk SC, Pluta L, Bond JA & Recio L (1994) Molecular analysis of *lacI* mutants from bone marrow of B6C3F1 transgenic mice following inhalation exposure to 1,3-butadiene. Carcinogenesis, **15**(3): 471–477.

Sisk SC, Pluta LJ, Meyer KG, Wong BC & Recio L (1997) Assessment of the in vivo mutagenicity of ethylene oxide in the tissues of B6C3F1 *lacI* transgenic mice following inhalation exposure. Mutat Res, **391**: 153–164.

Skopek TR, Kort KL & Marino DR (1995) Relative sensitivity of the endogenous *Hprt* gene and *lacI* transgene in ENU-treated Big Blue® B6C3F1 mice. Environ Mol Mutagen, 26: 9–15.

Skopek TR, Kort KL, Marino DR, Mittal LV, Umbenhauer DR, Laws GM & Adams SP (1996) Mutagenic response of the endogenous *Hprt* gene and *lacI* transgene in benzo[*a*]pyrene-treated Big Blue® B6C3F1 mice. Environ Mol Mutagen, **28**: 376–384.

Slattery SD & Valentine CR (2003) Development of a microplate assay for the detection of single plaque-forming units of bacteriophage ΦX174 in crude lysates. Environ Mol Mutagen, **41**: 121–125.

Slikker W, Mei N & Chen T (2004) *N*-Ethyl-*N*-nitrosourea (ENU) increased brain mutations in prenatal and neonatal mice but not in the adults. Toxicol Sci, **81**(1): 112–120.

Stratagene (2002) λ *Select-cII*™ mutation detection system for Big Blue® rodents. Instruction manual. La Jolla, California, Stratagene, pp 1–24 (Catalogue No. 720120; http://www.stratagene.com/manuals/720120.pdf).

Stuart GR, Thorleifson E, Okochi E, de Boer JG, Ushijima T, Nagao M & Glickman BW (2000a) Interpretation of mutational spectra from different genes: analyses of PhIP-induced mutational specificity in the *lacI* and *cII* transgenes from colon of Big Blue® rats. Mutat Res, **452**: 101–121.

Stuart GR, Holcroft J, de Boer JG & Glickman BW (2000b) Prostate mutations in rats induced by the suspected human carcinogen 2-amino-1-methyl-6-phenylimidazo[4,5-*b*]pyridine. Cancer Res, **60**: 266–268.

Stuart GR, de Boer JG, Haesevoets R, Holcroft J, Kangas J, Sojonky K, Thorleifson E, Thornton A, Walsh DF, Yang H & Glickman BW (2001) Mutations induced by 2-amino-1-methyl-6-phenylimidazo [4,5-*b*]pyridine (PhIP) in cecum and proximal and distal colon of *lacI* transgenic rats. Mutagenesis, **16**(5): 431–437.

Styles JA & Penman MG (1985) The mouse spot test. Evaluation of its performance in identifying chemical mutagens and carcinogens. Mutat Res, **154**(3): 183–204.

Sun B & Heddle JA (1999) The relationship between mutant frequency and time in vivo: simple predictions for any tissue, cell type, or mutagen. Mutat Res, **425**: 179–183.

Suter W, Ahiabor R, Blanco B, Locher F, Mantovani F, Robinson M, Sreenan G, Staedtler F, Swingler T, Vignutelli A & Perentes E (1996) Evaluation of the in vivo genotoxic potential of three carcinogenic aromatic amines using the Big Blue® transgenic mouse mutation assay. Environ Mol Mutagen, **28**: 354–362.

Suzuki T, Hayashi M, Sofuni T & Myhr BC (1993) The concomitant detection of gene mutation and micronucleus induction by mitomycin C in vivo using *lacZ* transgenic mice. Mutat Res, **285**: 219–224.

Suzuki T, Hayashi M & Sofuni T (1994) Initial experiences and future directions for transgenic mouse mutation assays. Mutat Res, **307**: 489–494.

Suzuki T, Hayashi M, Myhr B & Sofuni T (1995) Diethylnitrosamine is mutagenic in liver but not in bone marrow of *lacZ* transgenic mice (Muta™Mouse). Honyu Dobutsu Shiken Bunkakai Kaiho, **3**(1): 33–39.

Suzuki T, Itoh T, Hayashi M, Nishikawa Y, Ikezaki S, Furukawa F, Takahashi M & Sofuni T (1996a) Organ variation in the mutagenicity of dimethylnitrosamine in Big Blue® mice. Environ Mol Mutagen, **28**: 348–353.

Suzuki T, Hayashi M, Ochiai M, Wakabayashi K, Ushijima T, Sugimura T, Nagao M & Sofuni T (1996b) Organ variation in the mutagenicity of MeIQ in Big Blue® *lacI* transgenic mice. Mutat Res, **369**: 45–49.

Suzuki T, Hayashi M, Wang X, Yamamoto K, Ono T, Myhr BC & Sofuni T (1997) A comparison of the genotoxicity of ethylnitrosourea and ethyl methanesulfonate in *lacZ* transgenic mice (Muta™Mouse). Mutat Res, **395**: 75–82.

Suzuki T, Miyata Y, Saeki K, Kawazoe Y, Hayashi M & Sofuni T (1998) In vivo mutagenesis by the hepatocarcinogen quinoline in the *lacZ* transgenic mouse: evidence for its in vivo genotoxicity. Mutat Res, **412**: 161–166.

Suzuki T, Itoh S, Nakajima M, Hachiya N & Hara T (1999a) Target organ and time-course in the mutagenicity of five carcinogens in Muta™Mouse: a summary report of the second collaborative study of the transgenic mouse mutation assay by the JEMS/MMS. Mutat Res, **444**: 259–268.

Suzuki T, Uno Y, Idehara K, Baba T, Maniwa J, Ohkouchi A, Wang X, Hayashi M, Sofuni T, Tsuruoka M, Miyajima H & Kondo K (1999b) Procarbazine genotoxicity in the Muta™Mouse; strong clastogenicity and organ-specific induction of *lacZ* mutations. Mutat Res, **444**: 269–281.

Swiger RR (2001) Just how does the *cII* selection system work in Muta Mouse? Environ Mol Mutagen, **37**(4): 290–296.

Swiger RR, Cosentino L, Shima N, Bielas JH, Cruz-Munoz W & Heddle JA (1999) The *cII* locus in the Muta™Mouse system. Environ Mol Mutagen, **34**: 201–207.

Sykes PJ, Hooker AM, Jacobs AK, Harrington CS, Kingsbury L & Morley AA (1998) Induction of somatic intrachromosomal recombination inversion events by cyclophosphamide in a transgenic mouse model. Mutat Res, **397**: 209–219.

Sykes PJ, Hooker AM & Morley AA (1999) Inversion due to intrachromosomal recombination produced by carcinogens in a transgenic mouse model. Mutat Res, **427**: 1–9.

Sykes PJ, McCallum BD, Bangay MJ, Hooker AM & Morley AA (2001) Effect of 900 MHz radiofrequency radiation exposure on intrachromosomal recombination in pKZ1 mice. Radiat Res, **156**: 495–502.

Takeiri A, Mishima M, Tanaka K, Shioda A, Ueda O, Suzuki H, Inoue M, Masumura K & Nohmi T (2003) Molecular characterization of mitomycin C-induced large deletions and tandem-base substitution in the bone marrow of *gpt* delta transgenic mice. Chem Res Toxicol, **16**: 171–179.

Tao KS, Urlando C & Heddle JA (1993a) Comparison of somatic mutations in a transgenic versus host locus. Proc Natl Acad Sci USA, **90**: 10681–10685.

Tao KS, Urlando C & Heddle JA (1993b) Mutagenicity of methyl methanesulfonate (MMS) in vivo at the *Dlb-1* native locus and a *lacI* transgene. Environ Mol Mutagen, **22**: 293–296.

Thompson ED & Osterhues MA (1995) Effect of expression period on dimethylbenzanthracene-induced mutations in skin of Big Blue® mice. Abstr Annu Meet Environ Mutagen Soc, **1995**: 52.

Thornton AS, Oda Y, Stuart GR, Glickman BW & de Boer JG (2001) Mutagenicity of TCDD in Big Blue transgenic rats. Mutat Res, **478**(1–2): 45–50.

Thybaud V, Dean S, Nohmi T, de Boer J, Douglas GR, Glickman BW, Gorelick NJ, Heddle JA, Heflich RH, Lambert I, Martus HJ, Mirsalis JC, Suzuki T & Yajima N (2003) In vivo transgenic mutation assays. Mutat Res, **540**(2): 141–151.

Tice RR, Agurell E, Anderson D, Burlinson B, Hartmann A, Kobayashi H, Miyamae Y, Rojas E, Ryu J-C & Sasaki YF (2000) Single cell gel/comet assay: Guidelines for in vitro and in vivo genetic toxicology testing. Environ Mol Mutagen, **35**: 206–221.

Tinwell H, Lefevre PA & Ashby J (1994a) Response of the Muta™Mouse *lacZ*/*galE*$^-$ transgenic mutation assay to DMN: Comparison with the corresponding Big Blue® (*lacI*) responses. Mutat Res, **307**: 169–173.

Tinwell H, Lefevre PA & Ashby J (1994b) Mutation studies with dimethyl nitrosamine in young and old *lacI* transgenic mice. Mutat Res, **307**: 501–508.

Tinwell H, Lefevre PA & Ashby J (1998) Relative activities of methyl methanesulphonate (MMS) as a genotoxin, clastogen and gene mutagen to the liver and bone marrow of Muta™Mouse mice. Environ Mol Mutagen, **32**: 163–172.

Tombolan F, Renault D, Brault D, Guffroy M, Périn F & Thybaud V (1999a) Effect of mitogenic or regenerative cell proliferation on *lacZ* mutant frequency in the liver of Muta™Mice treated with 5,9-dimethyldibenzo[*c*,*g*]carbazole. Carcinogenesis, **20**: 1357–1362.

Tombolan F, Renault D, Brault D, Guffroy M, Perin-Roussel O, Perin F & Thybaud V (1999b) Kinetics of induction of DNA adducts, cell proliferation and gene mutations in the liver of Muta™Mice treated with 5,9-dimethyldibenzo[*c*,*g*]carbazole. Carcinogenesis, **20**(1): 125–132.

Topinka J, Oesterle D, Reimann R & Wolff T (2004a) No-effect level in the mutagenic activity of the drug cyproterone acetate in rat liver. Part I: Single dose treatment. Mutat Res, **550**(1–2): 89–99.

Topinka J, Oesterle D, Reimann R & Wolff T (2004b) No-effect level in the mutagenic activity of the drug cyproterone acetate in rat liver. Part II: Multiple dose treatment. Mutat Res, **550**(1–2): 101–108.

Topinka J, Loli P, Georgiadis P, Dusinska M, Hurbankova M, Kovacikova Z, Volkovova K, Kazimirova A, Barancokova M, Tatrai E, Oesterle D, Wolff T & Kyrtopoulos SA (2004c) Mutagenesis by asbestos in the lung of λ-*lacI* transgenic rats. Mutat Res, **553**(1–2): 67–78.

Trzos RJ, Petzold GL, Brunden MN & Swenberg JA (1978) The evaluation of sixteen carcinogens in the rat using the micronucleus test. Mutat Res, **58**: 79–86.

Tsutsui Y, Tsukidate K & Hakura A (1999) Sequence of the *lacZ* transgene of the Muta™Mouse. Environ Mutagen Res Commun, **21**: 39–43.

Tucker JD, Carrano AV, Allen NA, Christensen ML, Knize MG, Strout CL & Felton JS (1989) In vivo cytogenetic effects of cooked food mutagens. Mutat Res, **224**(1): 105–113.

Turner SD, Tinwell H, Piegorsch W, Schmezer P & Ashby J (2001) The male rat carcinogens limonene and sodium saccharin are not mutagenic to male Big Blue rats. Mutagenesis, **16**(4): 329–332.

Tweats DJ & Gatehouse DG (1988) Further debate of testing strategies. Mutagenesis, **3**(2): 95–102.

Ushijima T, Hosoya Y, Ochiai M, Kushida H, Wakabayashi K, Suzuki T, Hayashi M, Sofuni T, Sugimura T & Nagao M (1994) Tissue-specific mutational spectra of 2-amino-3,4-dimethylimidazo[4,5-*f*]quinoline in the liver and bone marrow of *lacI* transgenic mice. Carcinogenesis, **15**: 2805–2809.

Valentine CR, Montgomery BA, Miller SG, Delongchamp RR, Fane BA & Malling HV (2002) Characterization of mutant spectra generated by a forward mutational assay for gene *A* of ΦX174 from ENU-treated transgenic mouse embryonic cell line PX-2. Environ Mol Mutagen, **39**: 55–68.

Valentine CR, Raney JL, Shaddock JG, Dobrovolsky VN & Delongchamp RR (2004) In vivo mutation in gene *A* of splenic lymphocytes from ΦX174 transgenic mice. Environ Mol Mutagen, **44**: 128–150.

van Delft JH, Bergmans A, van Dam FJ, Tates AD, Howard L, Winton DJ & Baan RA (1998) Gene-mutation assay in lambda *lacZ* transgenic mice: comparison of *lacZ* with endogenous genes in splenocytes and small intestinal epithelium. Mutat Res, **415**: 85–96.

van Oostrom CT, Boeve M, van Den Berg J, de Vries A, Dolle ME, Beems RB, van Kreijl CF, Vijg J & van Steeg H (1999) Effect of heterozygous loss of *p53* on benzo[*a*]pyrene-induced mutations and tumors in DNA repair-deficient XPA mice. Environ Mol Mutagen, **34**: 124–130.

van Sloun PPH, Wijnhoven SWP, Kool HJM, Slater R, Weeda G, Van Zeeland AA, Lohman PHM & Vrieling H (1998) Determination of spontaneous loss of heterozygosity mutations in *Aprt* heterozygous mice. Nucleic Acids Res, **26**(21): 4888–4894.

Vijayalaxmi KK & Rai SP (1996) Studies on the genotoxicity of tamoxifen citrate in mouse bone marrow cells. Mutat Res, **368**(2): 109–114.

Vijg J & Douglas GR (1996) Bacteriophage lambda and plasmid *lacZ* transgenic mice for studying mutations in vivo. In: Pfeifer GP ed. Technologies for detection of DNA damage and mutations. New York, Plenum Press, pp 391–410.

Vijg J, Dolle ME, Martus HJ & Boerrigter ME (1997) Transgenic mouse models for studying mutations in vivo: applications in aging research. Mech Ageing Dev, **99**: 257–271.

Vogel U, Thein N, Moller P & Wallin H (2001) Pharmacological coal tar induces G:C to T:A transversion mutations in the skin of Muta Mouse. Pharmacol Toxicol, **89**(1): 30–34.

von Pressentin MdM, Kosinska W & Guttenplan JB (1999) Mutagenesis induced by oral carcinogens in *lacZ* mouse (Muta™Mouse) tongue and other oral tissues. Carcinogenesis, **20**: 2167–2170.

Wahnschaffe U, Bitsch A, Kielhorn J & Mangelsdorf I (2005a) Mutagenicity testing with transgenic mice. Part I: Comparison with the mouse bone marrow micronucleus test. J Carcinogen, **4**(3): 1–14.

Wahnschaffe U, Bitsch A, Kielhorn J & Mangelsdorf I (2005b) Mutagenicity testing with transgenic mice. Part II: Comparison with the mouse spot test. J Carcinogen, **4**(4): 1–8.

Walker VE, Gorelick NJ, O'Kelly JA, Craft TR, de Boer J, Glickman BW & Skopek TR (1994) Frequency and spectrum of ethylnitrosourea induced mutation at the *Hprt* and *lacI* loci in splenic T-cells of exposed Big Blue® mice. Environ Mol Mutagen Suppl, **23**: 71.

Walker VE, Sisk SC, Upton PB, Wong BA & Recio L (1997) In vivo mutagenicity of ethylene oxide at the *hprt* locus in T-lymphocytes of B6C3F1 *lacI* transgenic mice following inhalation exposure. Mutat Res, **392**: 211–222.

Walker VE, Andrews JL, Upton PB, Skopek TR, deBoer JG, Walker DM, Shi X, Sussman HE & Gorelick NJ (1999a) Detection of cyclophosphamide-induced mutations at the *Hprt* but not the *lacI* locus in splenic lymphocytes of exposed mice. Environ Mol Mutagen, **34**: 167–181.

Walker VE, Jones IM, Crippen TL, Meng Q, Walker DM, Bauer MJ, Reilly AA, Tates AD, Nakamura J, Upton PB & Skopek TR (1999b) Relationship between exposure, cell loss and proliferation, and manifestation of *Hprt* mutant T cells following treatment of preweanling, weanling, and adult male mice with *N*-ethyl-*N*-nitrosourea. Mutat Res, **431**: 371–388.

Wang J, Liu X, Heflich RH & Chen T (2004) Time course of *cII* gene mutant manifestation in the liver, spleen, and bone marrow of *N*-ethyl-*N*-nitrosourea-treated Big Blue® transgenic mice. Toxicol Sci, **82**(1): 124–128.

Wang X, Suzuki T, Itoh T, Honma M, Nishikawa A, Furukawa F, Takahashi M, Hayashi M, Kato T & Sofuni T (1998) Specific mutational spectrum of dimethylnitrosamine in the *lacI* transgene of Big Blue® C57BL/6 mice. Mutagenesis, **13**: 625–630.

Wang Y & Heddle JA (2004) Spontaneous and induced chromosomal damage and mutations in Bloom Syndrome mice. Mutat Res, **554**(1–2): 131–137.

Wijnhoven SW & van Steeg H (2003) Transgenic and knockout mice for DNA repair functions in carcinogenesis and mutagenesis. Toxicology, **193**: 171–187.

Wild D, Gocke E, Harnasch D, Kaiser G & King MT (1985) Differential mutagenic activity of IQ (2-amino-3-methylimidazo[4,5-*f*]quinoline) in *Salmonella typhimurium* strains in vitro and in vivo, in *Drosophila*, and in mice. Mutat Res, **156**: 93–102.

Williams CV, Fletcher K, Tinwell H & Ashby J (1998) Mutagenicity of ethyl carbamate to *lacZ*⁻ transgenic mice. Mutagenesis, **13**: 133–137.

Winegar RA, Carr G & Mirasalis JC (1997) Analysis of the mutagenic potential of ENU and MMS in germ cells of male C57BL/6 *lacI* transgenic mice. Mutat Res, **388**: 175–178.

Winton DJ, Blount MA & Ponder BAJ (1988) A clonal marker induced by mutation in mouse intestinal epithelium. Nature, **333**: 463–466.

Winton DJ, Gooderham NJ, Boobos AR, Davies DS & Ponder BAJ (1990) Mutagenesis of mouse intestine in vivo using the *Dlb-1* specific locus test: studies with 1,2-dimethylhydrazine, dimethylnitrosamine, and the dietary mutagen 2-amino-3,8-dimethylimidazo[4,5-*f*]quinoxaline. Cancer Res, **50**: 7992–7996.

Wolff T, Topinka J, Deml E, Oesterle D & Schwarz LR (2001) Dose dependent induction of DNA adducts, gene mutations and cell proliferation by the antiandrogenic drug cyproterone acetate in rat liver. Adv Exp Med Biol, **500**: 678–696.

Yamada K, Suzuki T, Kohara A, Hayashi M, Mizutani T & Saeki K (2004) In vivo mutagenicity of benzo[*f*]quinoline, benzo[*h*]quinoline, and 1,7-phenanthroline using the *lacZ* transgenic mice. Mutat Res, **559**(1–2): 83–95.

Yang H, Glickman B & de Boer JG (2002) Sex-specific induction of mutations by PhIP in the kidney of male and female rats and its modulations by conjugated linoleic acid. Environ Mol Mutagen, **40**(2): 116–121.

Yatagai F, Kurobe T, Nohmi T, Masumura K, Tsukada T, Yamaguchi H, Kasai-Eguchi K & Fukunishi N (2002) Heavy-ion-induced mutations in the *gpt* delta transgenic mouse: Effect of *p53* gene knockout. Environ Mol Mutagen, **40**: 216–225.

Zhang XB, Felton JS, Tucker JD, Urlando C & Heddle JA (1996) Intestinal mutagenicity of two carcinogenic food mutagens in transgenic mice: 2-amino-1-methyl-6-phenylimidazo[4,5-*b*]pyridine and amino(α)carboline. Carcinogenesis, **17**: 2259–2265.

Zimmer DM, Harbach PR, Mattes WB & Aaron CS (1999) Comparison of mutant frequencies at the transgenic λ *LacI* and *cII/cI* loci in control and ENU-treated Big Blue® mice. Environ Mol Mutagen, **33**: 249–256.

APPENDIX 1: MASTER TABLE

Substances were chosen for inclusion in the Master Table (Table A1-1) for the following reasons:

1) Substances for which there were transgenic studies carried out on more than one target organ and for which data on carcinogenicity are available were selected.
2) Substances for which data on Big Blue® mouse or Muta™Mouse *and* data on the mouse spot test or the mouse bone marrow micronucleus test were available were included.
3) Substances with negative results in carcinogenicity studies (and for which data on transgenic animals are available) were added.
4) Non-genotoxic carcinogens with available data on transgenic mutagenesis were added.

Master Table Legend

m: male; f: female; b: both genders; +: positive; (+): weak positive; −: negative; ±: inconclusive; bm: bone marrow; conver.: conversion; i.p.: intraperitoneal; nd: no data; mamm. cells: mammalian cells; p: peripheral blood reticulocytes used in the micronucleus test in vivo; s.c.: subcutaneous; transf.: transformation; 6-BT: 6-(*p*-dimethyl-aminophenylazo)-benzothiazole; CAS: Chemical Abstracts Service registry number; CNS: central nervous system; MTD: maximum tolerated dose; Na/K-ATPase: sodium/potassium adenosine triphosphatase; NTP: United States National Toxicology Program; PNS: peripheral nervous system; SLRL: sex-linked recessive lethal; SCE: sister chromatid exchange: UDS: unscheduled DNA synthesis

\# Study with negative results but limited validity (compare with data in Table 18).

[a] Data on carcinogenicity *and* genotoxicity are taken from sources (almost all secondary literature) cited in this column. The used sources were, for example, documents prepared by IARC, the German MAK Commission and WHO/IPCS (EHCs or CICADs). Data banks like HSDB, CCRIS, IRIS and GENE-TOX were also used if none of the above assessment documents was available for

a particular substance or to update the available information if the documents were prepared before the year 2000.

Data on the mouse spot test and the bone marrow micronucleus test in vivo are taken from reviews on these test systems or primary literature (see numeration in column "gene mutation in vivo" and "other genotoxic end-points in vivo").

[b] Predominantly carcinogenicity data on mice are presented in the case where transgenic mutation data are available on this species; if no data on mice are available or additional information on target organs is available in studies on rats, then data on rats are tabulated; if data on transgenic rat mutagenicity assays are available (e.g. sodium saccharin), then predominantly rat data are presented for carcinogenic effects.

[c] Ranking of target organs related to incidence.

[d] All available data (primary literature) are presented (exception *N*-ethyl-*N*-nitrosourea, also used as positive control; only selected studies presented to avoid repetition); ranking of organs according to mutagenic potency in the same study (same numerical superscript) related to controls of the same organ; results according to authors' evaluation.

Table A1-1. Comparison of target organs in long-term carcinogenicity bioassays with target organs of mutagenicity in the transgene mouse/rat assays and with target organs of mutagenicity in standard assays

Substance[a] (IARC classification)	Carcinogenicity		Transgenic animal models[d]		Genotoxicity			
	Species,[b] strain, sex, route	Target organs,[c] sex with positive results	Muta™Mouse sex, route, tested organ, result	Big Blue® species, strain, sex, route, tested organ, result	Gene mutation in vitro (test system)	Gene mutation in vivo (species)	Other genotoxic endpoints in vitro (test system)	Other genotoxic endpoints in vivo (species)
2-Acetylaminofluorene[1–3] CAS 53-96-3 (no evaluation)	Mouse, C57-C3H, nd, diet Rat, Slonaker, b, diet	Liver, nd Bladder, nd Kidney, nd Liver, b Bladder, b Kidney, b Acoustic duct, b	m, oral Liver[4] **+m**	Mouse, B6C3F1, m, i.p. Liver[5] **+m** Mouse, C57BL/6, f, oral Liver[1, 6] **+f**	Ames test **+** (bacteria) Mouse lymphoma **+** (mamm. cells) Gene mutation **+** (mamm. cells)	Spot test **+** (mouse)[7–9] SLRL **±** (*Drosophila*)	Cytogenetic **+** (mamm. cells) SCE **+** (mamm. cells) SCE **–** (human cells) UDS **±** (human cells) Cell transf. **±** (mamm. cells) gene conver. **±** (fungi)	Micronuclei **+** (rat[10] & mouse[11])

Table A1-1 (Contd)

Substance[a]	Carcinogenicity		Transgenic animal models[d]		Genotoxicity			
(IARC classification)	Species,[b] strain, sex, route	Target organs,[c] sex with positive results	Muta™Mouse sex, route, tested organ, result	Big Blue® species, strain, sex, route, tested organ, result	Gene mutation in vitro (test system)	Gene mutation in vivo (species)	Other genotoxic end-points in vitro (test system)	Other genotoxic end-points in vivo (species)
4-Acetyl-amino-fluorene[12–15] CAS 28322-02-3 (no evaluation)	Rat, nd, nd, nd	Insufficient study design, no conclusion	m, oral Liver[16] **+m**		Ames test **+** (bacteria) Mouse lymphoma **+** (mamm. cells)	Spot test **–** (mouse)[7, 8]	UDS **–** (mamm. cells) SCE **±** (non-human cells) Cell transf. **+** (mamm. cells) gene conver. **–** (fungi) DNA damage **±** (bacteria)	Micronuclei **±** (mouse)[10]

Table A1-1 (Contd)

Substance[a] (IARC classification)	Carcinogenicity		Transgenic animal models[d]		Genotoxicity			
	Species,[b] strain, sex, route	Target organs,[c] sex with positive results	Muta™Mouse sex, route, tested organ, result	Big Blue® species, strain, sex, route, tested organ, result	Gene mutation in vitro (test system)	Gene mutation in vivo (species)	Other genotoxic end-points in vitro (test system)	Other genotoxic end-points in vivo (species)
Acrylamide[17–19] CAS 79-06-1 (Group 2A)	Mouse, A/He, b, i.p.	Lung, b (limited validity; high incidence in control)	m, i.p. Bone marrow[20, 21] **+m** Liver[22] **(+)m** Testis[22] **±m**		Ames test **–** (bacteria) *Hprt* **+** (mamm. cells) Mouse lymphoma **+** (mamm. cells)	Spot test **+** (mouse)[23] Specific locus test **+** (mouse) Host-mediated assay **+** (mouse) Somatic mutation **+** (*Drosophila*) SLRL **+** (*Drosophila*)	Cytogenetic **+** (mamm. cells) SCE **+** (human & mamm. cells) Cell transf. **+** (mamm. cells) UDS **–** (mamm. cells) UDS **(+)** (human cells) Aneuploidy **+** (mamm. cells) DNA damage **+** (bacteria)	Micronuclei **±** (mouse)[17] Micronuclei **+** (not bm, mouse) Cytogenetic **+** (mouse) Dominant lethal **+** (mouse & rat) Heritable translocation **+** (mouse) SCE **+** (mouse) UDS **+** (rat) DNA damage **+** (mouse) DNA binding **+** (mouse & rat)
	Mouse, Swiss-ICR, f, gavage	Lung, f Skin, f (no further organs investigated)						
	Rat, F344, b, drinking-water	Testis, m Thyroid, b CNS, f Adrenal, m Oral cavity, f Uterus, f Clitoral gland, f Pituitary, f Mammary, f						

Table A1-1 (Contd)

Substance[a]	Carcinogenicity		Transgenic animal models[d]		Genotoxicity			
(IARC classification)	Species,[b] strain, sex, route	Target organs,[c] sex with positive results	Muta™Mouse sex, route, tested organ, result	Big Blue® species, strain, sex, route, tested organ, result	Gene mutation in vitro (test system)	Gene mutation in vivo (species)	Other genotoxic end-points in vitro (test system)	Other genotoxic end-points in vivo (species)
Aflatoxin B1[24] CAS 1162-65-8 (Group 1)	Mouse, A/He, f, i.p.	Lung, f (limited validity; high incidence in vehicle control)		Mouse, C57BL/6, m, gavage Liver[25] **+m** Kidney[25] **+m** Large intestine[25] **±m**	Ames test **+** (bacteria) *Hprt* **+** (mamm. cells) Mouse lymphoma **+** (mamm. cells) Ouabain **+** (mamm. cells) Other gene mutation **+** (mamm. cells) Gene mutation **+** (fungi)	Gene mutation at *p53* **+** (human) SLRL **+** (*Drosophila*) Host-mediated assay **+** (mouse) Somatic mutation **+** (*Drosophila*)	DNA adducts **+** (mamm. cells) DNA damage **+** (bacteria) DNA damage **+** (mamm. cells) UDS **+** (mamm. cells) Gene conver. **+** (fungi) SCE **+** (mamm. cells) Cytogenetic **+** (mamm. cells) Cell transf. **+** (mamm. cells)	Micronuclei **+** (rat[28] & mouse[10]) DNA adducts **+** (mouse) UDS **+** (rat) SCE **+** (mouse & rat) Cytogenetic **+** (mouse & rat) Dominant lethal (**±**) (mouse) Dominant lethal **+** (rat)
	Mouse, C57BL × CH3, b, i.p.	Liver, nd		Mouse, C57BL/6, nd, i.p. Liver[26] **–**				
	Rat, F344, b, gavage	Liver, b		Rat, F344, f, gavage Liver[27] **+f**				
				Rat, F344, m, i.p. Liver[26] **+m**				

Table A1-1 (Contd)

Substance[a] (IARC classification)	Carcinogenicity		Transgenic animal models[d]		Genotoxicity			
	Species,[b] strain, sex, route	Target organs,[c] sex with positive results	Muta™Mouse sex, route, tested organ, result	Big Blue® species, strain, sex, route, tested organ, result	Gene mutation in vitro (test system)	Gene mutation in vivo (species)	Other genotoxic end-points in vitro (test system)	Other genotoxic end-points in vivo (species)
Agaritine[29] CAS 2757-90-6 (Group 3)	Mouse, Swiss, b, drinking-water Mouse, Swiss, b, drinking-water	No increased tumour incidence Lung, b Blood vessels, b (fungal metabolite of agaritine used)		Mouse, C57BL/6, f, diet Kidney[30] **(+)f** Fore-stomach[30] **(+)f** Liver[30] **−f** Lung[30] **−f** Glandular stomach[30] **−f**	Ames test **(+)** (bacteria)	nd	nd	nd

Table A1-1 (Contd)

Substance[a] (IARC classification)	Carcinogenicity		Transgenic animal models[d]		Genotoxicity			
	Species,[b] strain, sex, route	Target organs,[c] sex with positive results	Muta™Mouse sex, route, tested organ, result	Big Blue® species, strain, sex, route, tested organ, result	Gene mutation in vitro (test system)	Gene mutation in vivo (species)	Other genotoxic end-points in vitro (test system)	Other genotoxic end-points in vivo (species)
4-Aminobiphenyl[17, 31–33] CAS 92-67-1 (Group 1)	Mouse, BALB/c, b, drinking-water	Blood vessels, b Bladder, m Liver, f	m, gavage Bladder[34] **+m** Liver[34] **+m** Bone marrow[34] **+m**		Ames test **+** (bacteria) *Hprt* **+** (mamm. cells)	Host-mediated assay **+** (bacteria & mouse)	DNA damage **+** (bacteria & mamm. cells) UDS **+** (mamm. cells) UDS **–** (human cells) Cell transf. **+** (mamm. cells) Mitotic conver./ recombination **+** (fungi)	Micronuclei **+** (mouse)[17] Micronuclei **–** (rat)[11] SCE **+** (hamster)
	Mouse, nd, b, oral	Liver, b						
	Mouse, B6C3F1, m, s.c. or i.p.	Liver, m						

Table A1-1 (Contd)

Substance[a] (IARC classification)	Carcinogenicity		Transgenic animal models[d]		Genotoxicity			
	Species,[b] strain, sex, route	Target organs,[c] sex with positive results	Muta™Mouse sex, route, tested organ, result	Big Blue® species, strain, sex, route, tested organ, result	Gene mutation in vitro (test system)	Gene mutation in vivo (species)	Other genotoxic endpoints in vitro (test system)	Other genotoxic endpoints in vivo (species)
2-Amino-3,4-dimethylimidazo[4,5-*f*]quinoline (MeIQ)[35, 36] CAS 77094-11-2 (Group 2B)	Mouse, C57BL/6, f, diet	Liver, f Caecum, f Colon, f Forestomach, f Small intestine, f Blood vessels, f		Mouse, C57BL/6, f, diet Colon[37] **+f** Bone marrow[37] **+f** Liver[37] **+f** Forestomach[37] **+f** Heart[37] **–f** (experimental design similar to the 1st cancer study)	Ames test **+** (bacteria) *Hprt* **+** (mamm. cells) Ouabain **–** (mamm. cells) Diphtheria toxin **+** (mamm. cells) Other gene mutation **+** (mamm. cells)	Somatic gene mutation **+** (*Drosophila*) Host-mediated **+** (bacteria & mouse)	Cytogenetic **(+)** (mamm. cells) Micronuclei **+** (mamm. cells) SCE **+** (mamm. cells) DNA damage **+** (bacteria & mamm. cells) UDS **+** (mamm. cells) DNA binding **+** (mamm. cells)	SCE **+** (mouse) DNA damage **+** (rat) DNA binding **+** (mouse & rat)
	Mouse, CDF1, b, diet	Forestomach, b Liver, f						

Table A1-1 (Contd)

Substance[a] (IARC classification)	Carcinogenicity		Transgenic animal models[d]		Genotoxicity			
	Species,[b] strain, sex, route	Target organs,[c] sex with positive results	Muta™Mouse sex, route, tested organ, result	Big Blue® species, strain, sex, route, tested organ, result	Gene mutation in vitro (test system)	Gene mutation in vivo (species)	Other genotoxic end-points in vitro (test system)	Other genotoxic end-points in vivo (species)
2-Amino-3,8-dimethyl-imidazo[4,5-*f*]quinoxaline (MeIQx)[38–41] CAS 77500-04-0 (Group 2B)	Mouse, CDF1, b, diet	Liver, b Lung, f Blood, lymphoma, m & leukaemia (no differentiation)		Mouse, C57BL/6, b, diet Liver[42] **+b** Colon[42] **+b**	Ames test **+** (bacteria) *Hprt* **±** (mamm. cells) Diphtheria toxin **+** (mamm. cells) Other gene mutation **+** (mamm. cells)	*Dlb-1* gene mutation **–** (mouse) Somatic gene mutation **+** (*Drosophila*) Host-mediated **+** (bacteria & mouse)	Cytogenetic **–** (human cells) Cytogenetic **+** (mamm. cells) SCE **+** (human & mamm. cells) UDS **+** (mamm. cells) DNA binding **+** (mamm. cells) DNA damage **+** (bacteria)	Micronuclei **–** (mouse)[40] Cytogenetic **+** (rat) SCE **+** (rat) SCE **–** (mouse) DNA binding **+** (rat & mouse)
	Rat, F344, b, diet	Liver, b Zymbal gland, b Clitoral gland, f Skin, m		Rat, F344, m, diet Liver[41] **+m** Colon[41] **+m** Zymbal gland[41] **+m** Kidney[41] **+m** Spleen[41] **(+)m** Lung[41] **–m** Testis[41] **–m** Heart[41] **–m** Brain[41] **–m** Fat tissue[41] **–m** Skeletal muscle[41] **–m**				

Table A1-1 (Contd)

Substance[a] (IARC classification)	Carcinogenicity		Transgenic animal models[d]		Genotoxicity			
	Species,[b] strain, sex, route	Target organs,[c] sex with positive results	Muta™Mouse sex, route, tested organ, result	Big Blue® species, strain, sex, route, tested organ, result	Gene mutation in vitro (test system)	Gene mutation in vivo (species)	Other genotoxic endpoints in vitro (test system)	Other genotoxic endpoints in vivo (species)
2-Amino-1-methyl-6-phenyl-imidazo[4,5-*b*]pyridine (PhIP)[43, 44] CAS 105650-23-5 (Group 2B)	Rat, F344, m, diet	Prostate gland, m	m, gavage Colon[45] **+m** Small intestine[45] **+m** Liver[45] **(+)m** Kidney[45] **–m**	Rat, F344, b, diet Colon[46, 47] **+b** Caecum[47] **+b**	Ames test **+** (bacteria) *Hprt* **+** (mamm. cells)	Host-mediated **+** (bacteria & mouse)	Cytogenetic **+** (mamm. cells) SCE **+** (mamm. cells) UDS **+** (mamm. cells) DNA damage **+** (mamm. cells)	Micronuclei **±** (mouse)[39] Cytogenetic **+** (mouse) SCE **+** (mouse) DNA binding **+** (rat & monkey)
	Rat, F344, b, diet	Colon, b (lower incidence in f) Mammary, f		Rat, F344, f, gavage Mammary[48] **+f**				
	Rat, Nagase, m, diet	Small intestine, m Colon, m Caecum, m		Rat, F344, b, diet Kidney[49] **+b**				
	Mouse, CDF1, b, diet	Blood, lymphoma, b Lung, nd		Rat, F344, m, diet Prostate[50, 51] **+m**				

Table A1-1 (Contd)

Substance[a] (IARC classification)	Carcinogenicity		Transgenic animal models[d]		Genotoxicity			
	Species,[b] strain, sex, route	Target organs,[c] sex with positive results	Muta™Mouse sex, route, tested organ, result	Big Blue® species, strain, sex, route, tested organ, result	Gene mutation in vitro (test system)	Gene mutation in vivo (species)	Other genotoxic end-points in vitro (test system)	Other genotoxic end-points in vivo (species)
2-Amino-3-methyl-imidazo[4,5-*f*]quinoline (IQ)[52] CAS 76180-96-6 (Group 2A)	Mouse, CDF1, b, diet Rat, F344, b, diet	Liver, b Lung, b Fore-stomach, b Zymbal gland, b Colon, b Liver, b Small intestine, b Skin, m Clitoral gland, f	nd, gavage Liver[53] **+**	Rat, F344, m, gavage Liver[54] **+m** Colon[54] **+m** Kidney[51, 54] **+m**	Ames test **+** (bacteria) *Hprt* **+** (mamm. cells) Ouabain **–** (mamm. cells) Diphtheria toxin **+** (mamm. cells) Other gene mutation **+** (mamm. cells)	Spot test **–** (mouse)[55] *Hprt* **+** (rat) SLRL **+** (*Drosophila*) Somatic gene mutation **+** (*Drosophila*) Host-mediated **+** (bacteria & mouse)	Cytogenetic **–** (mamm. cells) Cytogenetic **±** (human cells) Micronuclei **(+)** (human cells) SCE **+** (mamm. cells) DNA damage **+** (bacteria & mamm. cells) UDS **+** (mamm. cells)	Micronuclei – (mouse)[10, 55] Cytogenetic **–** (mouse) Cytogenetic **+** (rat) SCE **+** (mouse & rat) DNA damage **+** (rat) DNA binding **+** (mouse & rat)

Table A1-1 (Contd)

Substance[a] (IARC classification)	Carcinogenicity		Transgenic animal models[d]		Genotoxicity			
	Species,[b] strain, sex, route	Target organs,[c] sex with positive results	Muta™Mouse sex, route, tested organ, result	Big Blue® species, strain, sex, route, tested organ, result	Gene mutation in vitro (test system)	Gene mutation in vivo (species)	Other genotoxic end-points in vitro (test system)	Other genotoxic end-points in vivo (species)
ortho-Anisidine[56] CAS 90-04-0 (Group 2B)	Mouse B6C3F1, b, diet Rat, F344, b, diet	Bladder, b Bladder, b Kidney, m Thyroid, m		Mouse, B6C3F1, f, gavage Bladder[57] **+f** Liver[57] **−f**	Ames test **+** (bacteria) Mouse lymphoma **+** (mamm. cells)	Host-mediated assay **±** (mouse) SLRL **−** (*Drosophila*) Somatic gene mutation **+** (*Drosophila*)	Cytogenetic **+** (mamm. & human cells) SCE **+** (mamm. cells) UDS **−** (mamm. cells) DNA damage **+** (mamm. cells) DNA damage **(+)** (bacteria) Mitotic recombination **+** (fungi)	Micronuclei **−** (mouse)[17, 56] Micronuclei **−** (rat) DNA binding **−** (mouse) UDS **−** (rat) DNA damage **−** (rat)

Table A1-1 (Contd)

Substance[a] (IARC classification)	Carcinogenicity		Transgenic animal models[d]		Genotoxicity			
	Species,[b] strain, sex, route	Target organs,[c] sex with positive results	Muta™Mouse sex, route, tested organ, result	Big Blue® species, strain, sex, route, tested organ, result	Gene mutation in vitro (test system)	Gene mutation in vivo (species)	Other genotoxic end-points in vitro (test system)	Other genotoxic end-points in vivo (species)
Asbestos crocidolite[58–60] CAS 12001-28-4 (Group 1)	Mouse, NMRI, nd, i.p. Rat, nd, nd, inhalation Rat, Wistar, nd, intrapleural	Mesothelial tissue, nd Lung, nd Mesothelial tissue, nd		Mouse, C57BL/6, m, inhalation Lung[61] **+m** Rat, F344, m, intratracheal instillation of amosite Lung[62] **(+)m**	Ames test **–** (bacteria) Gene mutation **–** (mamm. cells)	nd	SCE **–** (human cells) Cytogenetic **±** (human cells) SCE **+** (mamm. cells) Cytogenetic **+** (mamm. cells) Micronuclei **+** (mamm. cells) DNA damage **–** (mamm. cells) Cell transf. **+** (mamm. cells)	Micronuclei **–** (mouse, i.p., chrysotile)[17]

Table A1-1 (Contd)

Substance[a] (IARC classification)	Carcinogenicity		Transgenic animal models[d]		Genotoxicity			
	Species,[b] strain, sex, route	Target organs,[c] sex with positive results	Muta™Mouse sex, route, tested organ, result	Big Blue® species, strain, sex, route, tested organ, result	Gene mutation in vitro (test system)	Gene mutation in vivo (species)	Other genotoxic end-points in vitro (test system)	Other genotoxic end-points in vivo (species)
Benzene[63, 64] CAS 71-43-2 (Group 1)	Mouse, C57BL/6, b, inhalation	Blood, lymphoma, b Blood, leukaemia, f Zymbal gland, m		Mouse, B6C3F1, m, gavage Bone marrow[65] **+m** Spleen[65] **+m** Lung[65] **−m**	Ames test **−** (bacteria) Mouse lymphoma **−** (mamm. cells) Gene mutation **±** (different mamm. test systems) Gene mutation **+** (fungi)	Somatic gene mutation **±** (*Drosophila*)	DNA damage **+** (bacteria) Gene conver. **+** (fungi) DNA damage **−** (mamm. cells) SCE **−** (mamm. cells) Micronuclei **−** (mamm. cells) Cytogenetic **±** (mamm. cells) Cell transf. **+** (mamm. cells) UDS **−** (mamm. cells)	Micronuclei **+** (mouse[11, 68] & rat[10]) Mitotic recombination **+** (*Drosophila*) Cytogenetic **−** (*Drosophila*) Cytogenetic **+** (mouse, rat & human) SCE **+** (mouse & rat)
	Mouse, CD1, m, inhalation	Lung, m		Mouse, C57BL/6, m, inhalation Lung[66, 67] **+m** Spleen[67] **+m** Liver[67] **−m**				
	Mouse B6C3F1, b, gavage	Zymbal gland, m Mammary gland, f Blood, nd Adrenals, nd Ovary, f Liver, nd Lung, nd Preputial gland, m						

Table A1-1 (Contd)

Substance[a] (IARC classification)	Carcinogenicity		Transgenic animal models[d]		Genotoxicity			
	Species,[b] strain, sex, route	Target organs,[c] sex with positive results	Muta™Mouse sex, route, tested organ, result	Big Blue® species, strain, sex, route, tested organ, result	Gene mutation in vitro (test system)	Gene mutation in vivo (species)	Other genotoxic endpoints in vitro (test system)	Other genotoxic endpoints in vivo (species)
Benzo[*a*]-pyrene[69, 70] CAS 50-32-8 (Group 2A)	Mouse, A/J, f, gavage Mouse, CFW, f, diet Mouse, C57BL, nd, s.c. Mouse, ICR, b, i.p.	Stomach, f Lung, f Stomach, f Lung, f Thymus, f Blood, lymphoma & leukaemia, f Local sarcoma, nd Lung, b	m, gavage Colon[71, 72] **+m** Ileum[71] **+m** Fore-stomach[71, 72] **+m** Bone marrow[71] **+m** Spleen[71, 72] **+m** Glandular stomach[71, 72] **+m** Liver[71, 72] **+m** Lung[71, 72] **+m** Kidney[71] **+m** Heart[71] **+m** Brain[71] **−m**	Mouse, B6C3F1, m, i.p. Spleen[76] **+m** Mouse, C57BL/6, nd, i.p. Spleen[77, 78] **+** Mouse, C57BL/6, m, i.p. Liver[79, 80] **+m**	Ames test **+** (bacteria) *Hprt* **+** (mamm. cells) Mouse lymphoma **+** (mamm. cells) Ouabain **+** (mamm. cells) Diphtheria toxin **+** (mamm. cells)	Spot test **+** (mouse)[81] SLRL **−** (*Drosophila*) Somatic gene mutation **+** (*Drosophila*)	DNA damage **+** (bacteria) DNA damage **+** (mamm. cells) UDS **+** (mamm. cells) DNA adducts **+** (mamm. cells) Cytogenetic **+** (mamm. cells) Micronuclei **+** (mamm. cells) SCE **+** (mamm. cells) Cell transf. **+** (mamm. cells) Mitotic conver./ recombination **±** (fungi)	Micronuclei **+** p (mouse & rat)[82] Micronuclei **+** (mouse & rat)[10, 11] Dominant lethal **+** (mouse) Cytogenetic **+** (mouse, hamster) Cytogenetic **−** (rat) SCE **+** (mouse, rat & hamster) UDS **−** (mouse & rat)

Table A1-1 (Contd)

Substance[a] (IARC classification)	Carcinogenicity		Transgenic animal models[d]		Genotoxicity			
	Species,[b] strain, sex, route	Target organs,[c] sex with positive results	Muta™Mouse sex, route, tested organ, result	Big Blue® species, strain, sex, route, tested organ, result	Gene mutation in vitro (test system)	Gene mutation in vivo (species)	Other genotoxic endpoints in vitro (test system)	Other genotoxic endpoints in vivo (species)
Benzo[*a*]pyrene *(contd)*	Mouse, B6C3F1, m, i.p.	Liver, m	m, i.p. Liver[73] **+m**					DNA adducts **+** (mouse, rat) Transformation **+** (hamster) DNA damage **+** (*Drosophila*)
	Rat, Sprague-Dawley, b, diet	Forestomach, b Oesophagus, b Larynx, b	f, gavage Colon[74] **+f** Mammary[74] **+f** Lung[74, 75] **+f** Liver[74] **+f** Kidney[74, 75] **+f** Oral cavity[75] **+f**					
			f, diet/1× gavage Lung[74] **+f** Oral cavity[74] **+f** Tongue[74] **+f** Breast[74] **+f** Liver[74] **+f**					

Table A1-1 (Contd)

Substance[a] (IARC classification)	Carcinogenicity		Transgenic animal models[d]		Genotoxicity			
	Species,[b] strain, sex, route	Target organs,[c] sex with positive results	Muta™Mouse sex, route, tested organ, result	Big Blue® species, strain, sex, route, tested organ, result	Gene mutation in vitro (test system)	Gene mutation in vivo (species)	Other genotoxic end-points in vitro (test system)	Other genotoxic end-points in vivo (species)
Bromomethane (methyl bromide)[83–85] CAS 74-83-9 (Group 3)	Mouse, B6C3F1, b, inhalation Rat, Wistar, b, gavage Rat, Wistar, b, inhalation Rat, F344, b, diet	No neoplasia No neoplasia (re-evaluated) No neoplasia No neoplasia	m, gavage **#** Liver[86] **−m** Lung[86] **−m** Spleen[86] **−m** Bone marrow[86] **−m** Stomach[86] **−m** DNA damage in the liver even at lower doses		Ames test **+** (bacteria) Gene mutation **+** (bacteria) Mouse lymphoma **+** (mamm. cells)	SLRL **+** (*Drosophila*)	Cytogenetic **+** (human cells) SCE **+** (human cells) Cell transf. **−** (mamm. cells) DNA damage **+** (bacteria) UDS **−** (human cells)	Micronuclei **+** (mouse & rat)[87] Micronuclei **±** (human) Cytogenetic **−** (rat) Dominant lethal **−** (rat) SCE **±** (mouse) UDS **−** (rat) Mitotic recombination **+** (*Drosophila*) DNA damage **+** (mouse & rat)[86]

Table A1-1 (Contd)

Substance[a] (IARC classification)	Carcinogenicity		Transgenic animal models[d]		Genotoxicity			
	Species,[b] strain, sex, route	Target organs,[c] sex with positive results	Muta™Mouse sex, route, tested organ, result	Big Blue® species, strain, sex, route, tested organ, result	Gene mutation in vitro (test system)	Gene mutation in vivo (species)	Other genotoxic end-points in vitro (test system)	Other genotoxic end-points in vivo (species)
1,3-Butadiene[88] CAS 106-99-0 (Group 2A)	Mouse, B6C3F1, b, inhalation	Heart (haemangiosarcoma), b Blood, lymphoma, b Lung, b Forestomach, b Harderian gland, b Liver, b Mammary gland, f Ovary, f	m, inhalation Lung[89] **+m** Bone marrow[89] **–m** Liver[89] **–m**	Mouse, B6C3F1, m, inhalation Bone marrow[89–91] **+m**	Ames test **+** (bacteria) Reverse mutation **–** (*E. coli*, bacteria) Mouse lymphoma **–** (mamm. cells)	Spot test **+** (mouse)[92] *HPRT* **+** (human) *Hprt* **+** (mouse) Somatic gene mutation/ recombination **–** (*Drosophila*) SLRL **–** (*Drosophila*)	SCE **+** (mamm. cells) DNA damage **–** (mamm. cells)	Micronuclei **+** (mouse)[68, 93] Micronuclei **–** (rat)[93] Cytogenetic **+** (human & mouse) Micronuclei **–** (human) DNA damage **+** (mouse & rat) UDS **–** (mouse & rat) SCE **+** (mouse) SCE **–** (rat) Dominant lethal **+** (mouse), **–** (rat)
	Mouse, B6C3F1, m, inhalation	Preputial gland, m Kidney, m						

Table A1-1 (Contd)

Substance[a] (IARC classification)	Carcinogenicity		Transgenic animal models[d]		Genotoxicity			
	Species,[b] strain, sex, route	Target organs,[c] sex with positive results	Muta™Mouse sex, route, tested organ, result	Big Blue® species, strain, sex, route, tested organ, result	Gene mutation in vitro (test system)	Gene mutation in vivo (species)	Other genotoxic end-points in vitro (test system)	Other genotoxic end-points in vivo (species)
1,3-Butadiene *(contd)*								Heritable translocation **+** (mouse) DNA adducts **+** (mouse & rat)
Chlorambucil[17, 94–97] CAS 305-03-3 (Group 1) alkylating agent	Mouse, Swiss, b, i.p. Mouse, BALB, b, gavage Rat, CD, b, i.p.	Lung, b Blood, lymphoma, b Ovary, f Blood, lymphoma, b Lung, b Mammary, f Blood, leukaemia & lymphoma, m	m, i.p. Bone marrow[20, 21] **+m** Liver[21] **+m** Testis[21] **+m**		Ames test **+** (bacteria) *Hprt* **+** (mamm. cells) Gene mutation **+** (fungi)	SLRL **+** (*Drosophila*)	Cytogenetic **+** (human cells) SCE **+** (human & mamm. cells) Mitotic conversion/ recombination (fungi) **+**	Micronuclei **+** (mouse)[10, 17] Micronuclei **+** (rat) Cytogenetic **+** (rat) Cytogenetic **±** (human) SCE **+** (human)

Table A1-1 (Contd)

Substance[a] (IARC classification)	Carcinogenicity: Species,[b] strain, sex, route	Carcinogenicity: Target organs,[c] sex with positive results	Transgenic animal models[d]: Muta™Mouse sex, route, tested organ, result	Transgenic animal models[d]: Big Blue® species, strain, sex, route, tested organ, result	Genotoxicity: Gene mutation in vitro (test system)	Genotoxicity: Gene mutation in vivo (species)	Genotoxicity: Other genotoxic endpoints in vitro (test system)	Genotoxicity: Other genotoxic endpoints in vivo (species)
Chloroform[98] CAS 67-66-3 (Group 2B) example non-genotoxic carcinogen MAK K4	Mouse, B6C3F1, b, gavage Mouse, B6C3F1, f, drinking-water Mouse, BDF1, b, inhalation	Liver, b Blood, lymphoma, f No increased tumour incidence Kidney, m Liver, f		Mouse B6C3F1, f, inhalation Liver[99] **–f**	Ames test **–** (bacteria) *Hprt* **±** (mamm. cells) Mouse lymphoma **+** (mamm. cells)	Host-mediated gene mutation **(+)** (mouse & *S. typhimurium*) SLRL **–** (*Drosophila*) Somatic mutation **–** (*Drosophila*)	Cytogenetic **–** (human cells) SCE **+** (human & mamm. cells) UDS **–** (human & mamm. cells) DNA damage **±** (mamm. cells) Cell transf. **–** (mamm. cells) DNA damage **–** (bacteria) Mitotic conver./ recombination **±** (fungi)	Micronuclei **+** (rat)[100] Micronuclei **(+)** (mouse)[68]; in 3 studies Micronuclei **–** (mouse)[11] Cytogenetic **±** (mouse) Cytogenetic **+** (rat & hamster) SCE **+** (mouse) DNA damage **–** (rat & mouse) DNA binding **–** (mouse) DNA binding **(+)** (rat)

Table A1-1 (Contd)

Substance[a] (IARC classification)	Carcinogenicity		Transgenic animal models[d]		Genotoxicity			
	Species,[b] strain, sex, route	Target organs,[c] sex with positive results	Muta™Mouse sex, route, tested organ, result	Big Blue® species, strain, sex, route, tested organ, result	Gene mutation in vitro (test system)	Gene mutation in vivo (species)	Other genotoxic end-points in vitro (test system)	Other genotoxic end-points in vivo (species)
Cyclophosphamide[101, 102] CAS 50-18-0 (Group 1)	Mouse, NMRI, f, s.c.	Mammary gland, f Ovary, f Lung, f Connective tissue (local sarcoma), f	i.p., m Bone marrow[20, 21] **+m**	Mouse B6C3F1, m, i.p. Lung[103] **+m** Bladder[103] **+m** Kidney[103] **−m** Bone marrow[103] **−m** Splenic T-lymphocytes[104] **−m**	Ames test **+** (bacteria) Mouse lymphoma **+** (mamm. cells) Gene mutation **+** (fungi)	Spot test **+** (mouse)[81, 106] Host-mediated gene mutation **+** (mouse & *S. typhimurium*) SLRL **+** (*Drosophila*) Somatic mutation **+** (*Drosophila*)	Cytogenetic **+** (human & mamm. cells & fungi) SCE **+** (human & mamm. cells) DNA damage **+** (human cells, fungi & bacteria) UDS **+** (mamm. cells) Cell transf. **+** (mamm. cells) Gene conver./recombination **+** (fungi)	Micronuclei **+** (rat[28] & mouse[68, 107])[11] Micronuclei **+** p (mouse)[10] Cytogenetic **+** (human, rat, mouse & hamster) SCE **+** (human & rodent) Dominant lethal **+** (mouse & rat) DNA damage **+** (rodents) DNA binding **+** (mouse) Heritable translocation **+** (*Drosophila*)
	Mouse, dd, b, i.p.	Lung, nd Liver, nd Testis, m Mammary gland, nd						
	Mouse, Swiss, b, i.p.	Lung, b Bladder, m		Mouse, C57BL/6, m, i.p. Liver[105] **+m** Testis[105] **−m** Spleen[105] **−m**				
	Rat, Sprague-Dawley, b, drinking-water	Bladder, b Blood (nd), b		Mouse, nd, i.p. Spleen[77, 78] **+**				

Table A1-1 (Contd)

Substance[a] (IARC classification)	Carcinogenicity		Transgenic animal models[d]		Genotoxicity			
	Species,[b] strain, sex, route	Target organs,[c] sex with positive results	Muta™Mouse sex, route, tested organ, result	Big Blue® species, strain, sex, route, tested organ, result	Gene mutation in vitro (test system)	Gene mutation in vivo (species)	Other genotoxic end-points in vitro (test system)	Other genotoxic end-points in vivo (species)
2,4-Diaminotoluene[108, 109] CAS 95-80-7 (Group 2B)	Mouse, B6C3F1, b, diet Rat, F344, b, diet	Liver, b Blood, lymphoma, f Liver, b Mammary gland, b Subcutaneous tissue, b		Mouse C57BL/6, b, gavage Liver[110] **+b** B6C3F1, m, diet Liver[110–112] **+m**	Ames test **+** (bacteria) Mouse lymphoma **+** (mamm. cells) Gene mutation **+** (mamm. cells)	SLRL **+** (*Drosophila*)	Cytogenetic **+** (mamm. cells) SCE **+** (mamm. cells) UDS **+** (mamm. cells) Cell transf.**+** (mamm. cells) DNA damage **+** (mamm. cells)	Micronuclei **+** (rat)[17] Micronuclei **–** (mouse)[17] UDS **+** (rat) SCE **+** (mouse) Dominant lethal **–** (mouse) DNA damage **+** (rat) DNA binding **+** (rat)

Table A1-1 (Contd)

Substance[a] (IARC classification)	Carcinogenicity		Transgenic animal models[d]		Genotoxicity			
	Species,[b] strain, sex, route	Target organs,[c] sex with positive results	Muta™Mouse sex, route, tested organ, result	Big Blue® species, strain, sex, route, tested organ, result	Gene mutation in vitro (test system)	Gene mutation in vivo (species)	Other genotoxic end-points in vitro (test system)	Other genotoxic end-points in vivo (species)
2,6-Diaminotoluene HCl[113] CAS 15481-70-6 (no evaluation)	Mouse, B6C3F1, b, diet	No increased tumour incidence (in contrast to the analogue 2,4-diaminotoluene; NTP assays, MTD reached)		Mouse, B6C3F1, m, diet **#** Liver[110, 111] **−m**	Ames test **+** (bacteria) Mouse lymphoma **+** (mamm. cells)	SLRL **−** (*Drosophila*)	Micronuclei **+** (mamm. cells) Cytogenetic **+** (mamm. cells) SCE **−** (mamm. cells)	Micronuclei **+** (mouse)[68, 114] Cytogenetic **−** (rodent)
	Rat, F344, b, diet	No increased tumour incidence, see remarks on mouse study						

Table A1-1 (Contd)

Substance[a] (IARC classification)	Carcinogenicity		Transgenic animal models[d]		Genotoxicity			
	Species,[b] strain, sex, route	Target organs,[c] sex with positive results	Muta™Mouse sex, route, tested organ, result	Big Blue® species, strain, sex, route, tested organ, result	Gene mutation in vitro (test system)	Gene mutation in vivo (species)	Other genotoxic end-points in vitro (test system)	Other genotoxic end-points in vivo (species)
1,2-Dibromo-ethane[115] CAS 106-93-4 (Group 2A)	Mouse, B6C3F1, b, gavage Mouse, B6C3F1, b, inhalation	Fore-stomach, b Lung, b Lung, b Blood vessels, f Subcutaneous tissue, f Nasal cavity, f Mammary, f	m, i.p. Liver[116] **–m** Testis[116] **(+)m** m, inhalation Nasal cavity[117] **+m** Lung[117] **–m**		Ames test **+** (bacteria) Gene mutation **+** (bacteria) *HPRT* **+** (human cells) Mouse lymphoma **+** (mamm. cells) Gene mutation **+** (fungi, mamm. & human cells)	SLRL **+** (*Drosophila*) Somatic mutation **+** (*Drosophila*)	Micronuclei **+** (human cells) Cytogenetic **+** (mamm. cells) UDS **+** (mamm. cells) SCE **+** (human & mamm. cells) DNA damage **+** (bacteria & mamm. cells) Cell transf.**+** (mamm. cells)	Micronuclei **–** (mouse)[118] Cytogenetic **–** (mouse) Dominant lethal **–** (rat & mouse) UDS **+** (rat) SCE **(+)** (mouse) DNA damage **+** (rat & mouse) DNA binding **+** (rat & mouse)

Table A1-1 (Contd)

Substance[a]	Carcinogenicity		Transgenic animal models[d]		Genotoxicity			
(IARC classification)	Species,[b] strain, sex, route	Target organs,[c] sex with positive results	Muta™Mouse sex, route, tested organ, result	Big Blue® species, strain, sex, route, tested organ, result	Gene mutation in vitro (test system)	Gene mutation in vivo (species)	Other genotoxic end-points in vitro (test system)	Other genotoxic end-points in vivo (species)
1,2-Dibromo-3-chloropropane[119] CAS 96-12-8 (Group 2B)	Mouse, nd, b, oral Mouse, B6C3F1, b, inhalation	Fore-stomach, b Lung, b Nasal cavity, b	m, i.p. Liver[116] **–m** Testis[116] **(+)m**		Ames test **+** (bacteria) *Hprt* **(+)** (mamm. cells) Mouse lymphoma **+** (mamm. cells)	Spot test **+** (mouse)[120] Specific locus test **–** (mouse) SLRL **+** (*Drosophila*) Somatic mutation **+** (*Drosophila*)	Cytogenetic **+** (human & mamm. cells) UDS **+** (mamm. cells) SCE **+** (mamm. cells) DNA damage **+** (mamm. cells) Cell transf.**+** (mamm. cells)	Micronuclei **+** (mouse)[17, 121, 122] Micronuclei **+** (rat) Cytogenetic **+** (rat) Dominant lethal **–** (mouse) Dominant lethal **+** (rat) Heritable translocation **±** (*Drosophila*) DNA binding **+** (rat)

Table A1-1 (Contd)

Substance[a] (IARC classification)	Carcinogenicity: Species,[b] strain, sex, route	Carcinogenicity: Target organs,[c] sex with positive results	Transgenic animal models[d]: Muta™Mouse sex, route, tested organ, result	Transgenic animal models[d]: Big Blue® species, strain, sex, route, tested organ, result	Genotoxicity: Gene mutation in vitro (test system)	Genotoxicity: Gene mutation in vivo (species)	Genotoxicity: Other genotoxic endpoints in vitro (test system)	Genotoxicity: Other genotoxic endpoints in vivo (species)
1,2-Dichloroethane[123, 124] CAS 107-06-2 (Group 2B)	Mouse, B6C3F1, b, gavage	Lung, b Blood, lymphoma, b Liver, m Mammary, f Uterus, f	m, gavage Liver[116] **−m** m, i.p. Liver[116] **−m** Testis[116] **−m**		Ames test **+** (bacteria) Gene mutation **+** (bacteria) *Hprt* **+** (mamm. cells) Gene mutation **+** (human cells) Gene mutation **−** (fungi)	Spot test **±** (mouse)[9] Host-mediated **−** (bacteria & mouse) SLRL **+** (*Drosophila*) Somatic mutation **+** (*Drosophila*)	Micronuclei **+** (human cells) UDS **+** (mamm. cells) DNA damage **+** (bacteria & mamm. cells) Aneuploidy **+** (fungi) Aneuploidy **(+)** (mamm. cells) Cell transf. **±** (mamm. cells) DNA binding **+** (mamm. cells)	Micronuclei **−** (mouse)[17] Dominant lethal **−** (mouse) SCE **+** (mouse) DNA damage **+** (mouse & rat) DNA binding **+** (mouse & rat)

Table A1-1 (Contd)

Substance[a] (IARC classification)	Carcinogenicity		Transgenic animal models[d]		Genotoxicity			
	Species,[b] strain, sex, route	Target organs,[c] sex with positive results	Muta™Mouse sex, route, tested organ, result	Big Blue® species, strain, sex, route, tested organ, result	Gene mutation in vitro (test system)	Gene mutation in vivo (species)	Other genotoxic endpoints in vitro (test system)	Other genotoxic endpoints in vivo (species)
Di-(2-ethylhexyl) phthalate (DEHP)[125] CAS 117-81-7 (Group 2B) MAK K4	Mouse, B6C3F1, b, diet Rat, F344, b, diet	Liver, b Liver, b Pancreas, m		Mouse C57BL/6, f, diet **#** Liver[6] **–f**	Ames test **–** (bacteria) *Hprt* **–** (mamm. cells) Mouse lymphoma **–** (mamm. cells) Na/K-ATPase **–** (mamm. cells) Gene mutation **±** (fungi)	Spot test **–** (mouse, but co-recombinogenic effects)[126] SLRL **–** (*Drosophila*) Somatic mutation **–** (*Drosophila*) Somatic mutation **(+)** (*Drosophila*)	Micronuclei **–** (mamm. cells) Cytogenetic **–** (human & mamm. cells) SCE **–** (human & mamm. cells) DNA damage **–** (mamm. cells) UDS **–** (human & mamm. cells) DNA binding **–** (mamm. cells) Cell transf.**+** (mamm. cells) Mitotic recombination **–** (fungi) DNA damage **–** (bacteria)	Micronuclei **–** (mouse)[17] Cytogenetic **–** (rat) Cytogenetic **+** (hamster) Dominant lethal **±** (mouse) UDS **–** (rat & mouse) DNA damage **–** (rat) DNA binding **–** (rat)

Table A1-1 (Contd)

Substance[a] (IARC classification)	Carcinogenicity		Transgenic animal models[d]		Genotoxicity			
	Species,[b] strain, sex, route	Target organs,[c] sex with positive results	Muta™Mouse sex, route, tested organ, result	Big Blue® species, strain, sex, route, tested organ, result	Gene mutation in vitro (test system)	Gene mutation in vivo (species)	Other genotoxic end-points in vitro (test system)	Other genotoxic end-points in vivo (species)
5-(*p*-Dimethyl-amino-phenylazo)-benzothiazole (5-BT)[127] CAS 18463-90-6 (no evaluation)	Rat, Sprague-Dawley, nd, diet	No increased tumour incidence compared with the analogue 6-BT (same experimental design, but limited bioassay)		Rat, F344, m, diet Liver[127] **+m** Rat, F344, m, gavage Liver[127] **+m**	Ames test **+** (bacteria)	nd	nd	UDS **+** (rat)

Table A1-1 (Contd)

Substance[a] (IARC classification)	Carcinogenicity		Transgenic animal models[d]		Genotoxicity			
	Species,[b] strain, sex, route	Target organs,[c] sex with positive results	Muta™Mouse sex, route, tested organ, result	Big Blue® species, strain, sex, route, tested organ, result	Gene mutation in vitro (test system)	Gene mutation in vivo (species)	Other genotoxic endpoints in vitro (test system)	Other genotoxic endpoints in vivo (species)
(+)-Limonene[169] CAS 5989-27-5 (Group 3) example non-genotoxic carcinogen	Rat, F344/N, b, gavage Mouse, B6C3F1, b, gavage	Kidney, m (alpha2u-globulin involved) No increased tumour incidence		Rat F344, m, diet **#** Liver[170] **–m** Kidney[170] **–m**	Ames test – (bacteria) Mouse lymphoma – (mamm. cells)	Spot test – (mouse; reduces the effect of ENU)[81]	Cytogenetic – (mamm. cells) Cell transf. – (mamm. cells)	nd

Table A1-1 (Contd)

Substance[a] (IARC classification)	Carcinogenicity		Transgenic animal models[d]		Genotoxicity			
	Species,[b] strain, sex, route	Target organs,[c] sex with positive results	Muta™Mouse sex, route, tested organ, result	Big Blue® species, strain, sex, route, tested organ, result	Gene mutation in vitro (test system)	Gene mutation in vivo (species)	Other genotoxic endpoints in vitro (test system)	Other genotoxic endpoints in vivo (species)
6-(*p*-Dimethylaminophenylazo)-benzothiazole (6BT)[127, 128] CAS 18463-85-9 (no evaluation)	Rat, Sprague-Dawley, nd, diet Rat, Sprague-Dawley, nd, gavage Rat, AP, nd, gavage	Liver (100% incidence), nd Liver, nd Liver, nd		Rat, F344, m, diet Liver[127] **+m** Rat, F344, m, gavage Liver[127, 128] **+m**	Ames test **+** (bacteria)	nd	nd	Micronuclei **+** (rat)[128] UDS **+** (rat)

Table A1-1 (Contd)

Substance[a] (IARC classification)	Carcinogenicity		Transgenic animal models[d]		Genotoxicity			
	Species,[b] strain, sex, route	Target organs,[c] sex with positive results	Muta™Mouse sex, route, tested organ, result	Big Blue® species, strain, sex, route, tested organ, result	Gene mutation in vitro (test system)	Gene mutation in vivo (species)	Other genotoxic endpoints in vitro (test system)	Other genotoxic endpoints in vivo (species)
7,12-Dimethyl-benz[*a*]-anthracene[129, 130] CAS 57-97-6 (no evaluation)	Mouse, C57BL, f, dermal	Skin, f Mammary, f Ovary, f	m, dermal Skin[20, 21, 131] **+m** Bone marrow[21] **+m**	Mouse C57BL/6, m, dermal Skin[134] **+m**	Ames test **+** (bacteria) *Hprt* **+** (mamm. cells) Mouse lymphoma **+** (mamm. cells) Other gene mutation **+** (mamm. cells) Gene mutation **+** (plant)	*Hprt* **+** (mouse) Specific locus test **±** (mouse) Host-mediated **+** (bacteria/ rodent) SLRL **±** (*Drosophila*)	Cytogenetic **+** (mamm. cells) SCE **+** (human & mamm. cells) UDS **+** (mamm. cells) Cell transf. **+** (mamm. cells) gene conver./ recombination **+** (fungi) DNA damage **±** (bacteria)	Micronuclei **+** (mouse & rat)[10] Cytogenetic **+** (rodent) SCE **+** (rodent) UDS **+** (mouse)
	Mouse, BALB, f, gavage	Ovary, f		B6C3F1, nd, dermal Skin[135, 136] **+** Liver[135] **+** Lung[135] **+** Bone marrow[135] **+**				
	Mouse, Swiss, b, s.c.	Injection site, b Lung, f Blood, lymphoma, f	m, i.p. Bone marrow[132] **+m** Liver[132] **+m** Skin[132] **+m** Colon[132] **+m** Thymus[132] **+m** Kidney[132, 133] **+m** Testis[132, 133] **+m**					
	Mouse, CD-1, b, s.c.	Lung, b Liver, m		Rat, F344, f, gavage Bone marrow[137] **+f** Spleen[138] **+f** Mammary[139] **+f**				

Table A1-1 (Contd)

Substance[a] (IARC classification)	Carcinogenicity		Transgenic animal models[d]		Genotoxicity			
	Species,[b] strain, sex, route	Target organs,[c] sex with positive results	Muta™Mouse sex, route, tested organ, result	Big Blue® species, strain, sex, route, tested organ, result	Gene mutation in vitro (test system)	Gene mutation in vivo (species)	Other genotoxic endpoints in vitro (test system)	Other genotoxic endpoints in vivo (species)
7,12-Dimethylbenz[*a*]anthracene (*contd*)	Mouse, B6C3F1, m, i.p.	Liver, m	f, oral (swabbed) Pooled oral tissue[75] **+f** Tongue[75] **−f**					
	Rat, Wistar, f, gavage	Blood, leukaemia, f Mammary, f						

Table A1-1 (Contd)

Substance[a]	Carcinogenicity		Transgenic animal models[d]		Genotoxicity			
(IARC classification)	Species,[b] strain, sex, route	Target organs,[c] sex with positive results	Muta™Mouse sex, route, tested organ, result	Big Blue® species, strain, sex, route, tested organ, result	Gene mutation in vitro (test system)	Gene mutation in vivo (species)	Other genotoxic end-points in vitro (test system)	Other genotoxic end-points in vivo (species)
Ethylene oxide[140] CAS 75-21-8 (Group 1)	Mouse, B6C3F1, b, inhalation	Lung, b Harderian gland, b Blood, lymphoma, f Uterus, f Mammary gland, f		Mouse B6C3F1, m, inhalation Lung[141] **+m** Bone marrow[141] **−m** Bone	Ames test **+** (bacteria) Forward & reverse mutation **+** (*E. coli*, bacteria) *Hprt* **+** (mamm. cells)	*Hprt* **+** (mouse & hamster) *HPRT* **+** (human) SLRL **+** (*Drosophila*) Somatic	UDS **+** (mamm. cells) Micronuclei **+** (mamm. cells) Cytogenetic **+** (mamm. cells) SCE **+** (mamm. cells)	Micronuclei **+** (human, rat[11] & mouse[11, 144]) UDS **(+)** (human) Cytogenetic **+** (human, mouse & rat)

Table A1-1 (Contd)

Substance[a] (IARC classification)	Carcinogenicity		Transgenic animal models[d]		Genotoxicity			
	Species,[b] strain, sex, route	Target organs,[c] sex with positive results	Muta™Mouse sex, route, tested organ, result	Big Blue® species, strain, sex, route, tested organ, result	Gene mutation in vitro (test system)	Gene mutation in vivo (species)	Other genotoxic end-points in vitro (test system)	Other genotoxic end-points in vivo (species)
Ethylene oxide *(contd)*	Rat, F344, b, inhalation	Brain, b Blood, leukaemia, b Mesothelial tissue, m Connective tissue (fibroma), f		marrow[142] **+m** Testis[142] **+m** Spleen[141, 143] **−m** Germ cell[141] **−m**	Mouse lymphoma **+** (mamm. cells) Ouabain **+** (mamm. cells) Other gene mutation **+** (mamm. cells) Gene mutation **+** (fungi, plants)	mutation **+** (*Drosophila*)	Cell transf. **+** (mamm. cells) Gene conver. **+** (fungi) Cytogenetic **+** (plants)	SCE **+** (human, mouse & rat) Dominant lethal **+** (mouse & rat) Heritable translocation **+** (mouse & *Drosophila*) DNA damage **+** (mouse) DNA adducts **+** (mouse & rat)

Table A1-1 (Contd)

Substance[a] (IARC classification)	Carcinogenicity: Species,[b] strain, sex, route	Carcinogenicity: Target organs,[c] sex with positive results	Transgenic animal models[d]: Muta™Mouse sex, route, tested organ, result	Transgenic animal models[d]: Big Blue® species, strain, sex, route, tested organ, result	Genotoxicity: Gene mutation in vitro (test system)	Genotoxicity: Gene mutation in vivo (species)	Genotoxicity: Other genotoxic end-points in vitro (test system)	Genotoxicity: Other genotoxic end-points in vivo (species)
Ethyl methane-sulfonate[145–148] CAS 62-50-0 (Group 2B)	Mouse, CFW/D, b, i.p.	Lung, b	m, i.p. Bone marrow[149, 150] **+m**		Ames test **+** (bacteria) *Hprt* **+** (mamm. cells) Ouabain **+** (mamm. cells) Mouse lymphoma **+** (mamm. cells) Gene mutation **+** (fungi & plants)	Spot test **+** (mouse)[9, 81, 152] Specific locus test **+** (mouse) *Hprt*, host-mediated **+** (mamm. cells/ mouse) SLRL **+** (*Drosophila*)	Cytogenetic **+** (mamm. cells, fungi & plants) Micronuclei **+** (human & mamm. cells) SCE **+** (human & mamm. cells) UDS **+** (human & mamm. cells) Cell transf. **+** (mamm. cells) Gene conver./ recombination **+** (fungi) DNA damage **+** (bacteria, human & mamm. cells)	Micronuclei **+** (mouse & rat)[11] Cytogenetic **+** (mouse) UDS **+** (mouse) Dominant lethal **+** (rat, mouse & *Drosophila*) Heritable translocation **+** (mouse & *Drosophila*) SCE **+** (mouse & rat)
	Mouse, CBA, m, i.p.	Lung, m Kidney, m	Liver[149] **±m** Liver[149] **+m**					
	Mouse, ICR, f, s.c.	Thymus, lymphoma, f Lung, f	f, i.p. Bone marrow[151] **(+)f** Liver[151] **±f** Brain[151] **−f**					
	Rat, Porton, f, i.p.	Kidney, f						

Table A1-1 (Contd)

Substance[a] (IARC classification)	Carcinogenicity		Transgenic animal models[d]		Genotoxicity			
	Species,[b] strain, sex, route	Target organs,[c] sex with positive results	Muta™Mouse sex, route, tested organ, result	Big Blue® species, strain, sex, route, tested organ, result	Gene mutation in vitro (test system)	Gene mutation in vivo (species)	Other genotoxic end-points in vitro (test system)	Other genotoxic end-points in vivo (species)
N-Ethyl-*N*-nitrosourea (ENU)[153, 154] CAS 759-73-9 (Group 2A)	Mouse, C57BL, nd, s.c.	Liver, nd	m, i.p. Bone marrow[21, 89, 149, 155] **+m** Spleen[155] **+m** Bladder[155] **+m** Liver[21, 89, 149, 155] **+m** Lung[89, 155] **+m** Kidney[155] **+m** Heart[155] **+m** Brain[155] **−m** Testis[21] **+m** Germ cells[156] **+m**	Mouse C57BL/6, m, i.p. Germ cells[157–160] **+m** Spleen[161] **+m** Liver[161] **+m** Lung[161] **+m** B6C3F1, m, i.p. Spleen[162] **+m** Germ cells[163] **+m**	Ames test **+** (bacteria) *Hprt* **+** (mamm. cells) Gene mutation **+** (fungi & plants)	Spot test **+** (mouse)[9, 81, 106] *Hprt* **+** (mouse) Host-mediated gene mutation **+** (nd) SLRL **+** (*Drosophila*) Specific locus test **+** (mouse; gene mutation)	Cytogenetic **+** (mamm.& human cells) SCE **+** (mamm. & human cells) UDS **+** (mamm. & human cells) Cell transf. **+** (mamm. cells) Cytogenetic **+** (plants) Gene conver. **+** (fungi) DNA damage **+** (bacteria)	Micronuclei **+** (mouse)[10, 164] Cytogenetic **+** (rat & mouse) SCE **+** (non-human) UDS **+** (mouse) Cytogenetic **+** (*Drosophila*) Heritable translocation **+** (*Drosophila*) Cell transf. **+** (rat)
	Mouse, B6C3F1, b, i.p.	Liver, m Harderian gland, m Lymphoreticular, f Ovary, f Mammary gland, f						
	Rat, BD-IX, nd, gavage	Brain, nd PNS, nd						
	Rat, Dory, f, drinking-water	Blood, leukaemia, f	f, i.p. Bone marrow[151] **+f** Liver[73, 151] **+f** Brain[83, 90] **−f**					

Table A1-1 (Contd)

Substance[a] (IARC classification)	Carcinogenicity: Species,[b] strain, sex, route	Carcinogenicity: Target organs,[c] sex with positive results	Transgenic animal models[d]: Muta™Mouse sex, route, tested organ, result	Transgenic animal models[d]: Big Blue® species, strain, sex, route, tested organ, result	Genotoxicity: Gene mutation in vitro (test system)	Genotoxicity: Gene mutation in vivo (species)	Genotoxicity: Other genotoxic endpoints in vitro (test system)	Genotoxicity: Other genotoxic endpoints in vivo (species)
Heptachlor[165] CAS 76-44-8 (Group 2B) MAK K3B	Mouse, C3H, b, diet Mouse, B6C3F1, b, diet	Liver, b Liver, b		Mouse C57BL/6, f, diet **#** Liver[6] **–f**	Ames test **–** (bacteria) Mouse lymphoma **+** (mamm. cells) *Hprt* **–** (mamm. cells)	SLRL **–** (*Drosophila*)	UDS **–** (rodent cells) UDS **+** (human cells) DNA damage **–** (bacteria) Gene conver.**–** (fungi)	Dominant lethal **–** (mouse)

Table A1-1 (Contd)

Substance[a] (IARC classification)	Carcinogenicity		Transgenic animal models[d]		Genotoxicity			
	Species,[b] strain, sex, route	Target organs,[c] sex with positive results	Muta™Mouse sex, route, tested organ, result	Big Blue® species, strain, sex, route, tested organ, result	Gene mutation in vitro (test system)	Gene mutation in vivo (species)	Other genotoxic end-points in vitro (test system)	Other genotoxic end-points in vivo (species)
Hydrazine[166] CAS 302-01-2 Hydrazine sulfate CAS 10034-93-2 (Group 2B)	Mouse, Swiss, b, drinking-water Mouse, CBA, b, gavage Mouse, C57BL, f, inhalation	Lung, b Blood, lymphoma, m Liver, b Lung, f	m, gavage Lung[167] **−m** Liver[167] **−m** Bone marrow[167] **−m** single exposure		Ames test **+** (bacteria) Gene mutation **+** (bacteria) Mouse lymphoma **+** (mamm. cells) *Hprt* **−** (mamm. cells) Gene mutation **+** (human cells) Other gene mutation **±** (mamm. cells) Gene mutation **+** (fungi)	Spot test **+** (mouse)[168] Host-mediated **+** (bacteria & mouse) SLRL **+** (*Drosophila*) Somatic mutation **+** (*Drosophila*)	Cytogenetic **−** (human cells) Cytogenetic **+** (mamm. cells) UDS **+** (mamm. & human cells) SCE **+** (mamm. cells) DNA damage **+** (bacteria) DNA damage **−** (mamm. cells) Cell transf. **+** (mamm. cells) Gene conver./ recombination **+** (fungi)	Micronuclei **+** (after repeated dosing; mouse)[17, 166] Cytogenetic **−** (mouse) Cytogenetic **+** (*Drosophila*) Dominant lethal **−** (mouse) UDS **−** (mouse) SCE **−** (mouse) DNA damage **+** (mouse) negative results after single exposure

Table A1-1 (Contd)

Substance[a]	Carcinogenicity		Transgenic animal models[d]		Genotoxicity			
(IARC classification)	Species,[b] strain, sex, route	Target organs,[c] sex with positive results	Muta™Mouse sex, route, tested organ, result	Big Blue® species, strain, sex, route, tested organ, result	Gene mutation in vitro (test system)	Gene mutation in vivo (species)	Other genotoxic end-points in vitro (test system)	Other genotoxic end-points in vivo (species)
Methyl methane-sulfonate[171, 172] CAS 66-27-3 (Group 2B)	Mouse, RF/Un, m, drinking-water Rat, Sprague-Dawley, m, inhalation	Lung, m Thymus, lymphoma, m Nasal cavity, m	m, i.p. Liver[16] **(+)m** But induced micronuclei at the same dose[16] Bone marrow[16, 173] **−m** Germ cells[173–175] **−m** Testis[174] **−m** Spleen[174] **−m**	Mouse B6C3F1, m, i.p. Liver[176] **−m** C57BL/6, m, i.p. Germ cells[157–159] **−m**	Ames test **+** (bacteria) *HPRT* **+** (human & mamm. cells) Mouse lymphoma **+** (mamm. cells) Ouabain **+** (mamm. cells)	Spot test **+** (mouse)[79] *Hprt* **+** (rat) Specific locus test **±** (mouse) Somatic mutation **+** (*Drosophila*) SLRL **+** (*Drosophila*) Host-mediated gene mutation **+** (bacteria & mouse)	Cytogenetic **+** (mamm. cells) Micronuclei **+** (human & mamm. cells) UDS **+** (human & mamm. cells) SCE **+** (mamm. & human cells) DNA damage **+** (human & mamm. cells) Cell transf. **±** (mamm. cells)	Micronuclei **+** (rat[10] & mouse[11, 177]) Cytogenetic **+** (mouse) Dominant lethal **+** (mouse) Heritable translocation **+** (mouse) DNA damage **+** (mouse & rat) UDS **+** (mouse & rat)

Table A1-1 (Contd)

Substance[a] (IARC classification)	Carcinogenicity		Transgenic animal models[d]		Genotoxicity			
	Species,[b] strain, sex, route	Target organs,[c] sex with positive results	Muta™Mouse sex, route, tested organ, result	Big Blue® species, strain, sex, route, tested organ, result	Gene mutation in vitro (test system)	Gene mutation in vivo (species)	Other genotoxic end-points in vitro (test system)	Other genotoxic end-points in vivo (species)
N-Methyl-*N*'-nitro-*N*-nitroso-guanidine (MNNG)[178–180] CAS 70-25-7 (Group 2A)	Mouse, C3H, m, gavage Mouse, ICR, nd, dermal Rat, Wistar, m, drinking-water	Stomach, m Intestine, m Skin, nd Glandular stomach, m Intestine, m Fore-stomach, m Mesentery, m Liver, m	m, dermal Skin[20, 181] **+m** Stomach[181] **−m** m, gavage Stomach[181, 182] **+m** Liver[181, 182] **−m** Bone marrow[182] **−m**		Ames test **+** (bacteria) Forward mutation **+** (bacteria) *Hprt* **+** (mamm. cells) Mouse lymphoma **+** (mamm. cells) Other gene mutation **+** (mamm. cells) Gene mutation **+** (fungi & plants)	Spot test **+** (mouse)[81,152] Host-mediated gene mutation **+** (nd) Somatic mutation **+** (*Drosophila*) SLRL **+** (*Drosophila*)	Cytogenetic **+** (human & mamm. cells) SCE **+** (human & mamm. cells) UDS **+** (human & mamm. cells) Cell transf. **+** (mamm. cells) DNA damage **+** (human & mamm. cells, bacteria) Gene conver. **+** (fungi)	Micronuclei **+** (mouse & rat)[10] Cytogenetic **+** (mouse) SCE **+** UDS **+** (rat) Dominant lethal **−** (mouse)

Table A1-1 (Contd)

Substance[a] (IARC classification)	Carcinogenicity		Transgenic animal models[d]		Genotoxicity			
	Species,[b] strain, sex, route	Target organs,[c] sex with positive results	Muta™Mouse sex, route, tested organ, result	Big Blue® species, strain, sex, route, tested organ, result	Gene mutation in vitro (test system)	Gene mutation in vivo (species)	Other genotoxic end-points in vitro (test system)	Other genotoxic end-points in vivo (species)
4-(Methylnitrosamino)-1-(3-pyridyl)-1-butanone (NNK)[183, 184] CAS 64091-91-4 (Group 2B)	Mouse, Swiss, f, i.p. Mouse, C3H, f, i.p. Rat, F344, b, s.c.	Lung, f Liver, f Lung, f Nasal cavity, b Lung, b Liver, b	m, i.p. Liver[185] **+m** Lung[185] **+m**		Ames test **+** (bacteria) *Hprt* **+** (mamm. cells)	Host-mediated gene mutation **+** (bacteria & mouse)	UDS **+** (mamm. cells)	DNA adducts **+** (rat)

Table A1-1 (Contd)

Substance[a] (IARC classification)	Carcinogenicity		Transgenic animal models[d]		Genotoxicity			
	Species,[b] strain, sex, route	Target organs,[c] sex with positive results	Muta™Mouse sex, route, tested organ, result	Big Blue® species, strain, sex, route, tested organ, result	Gene mutation in vitro (test system)	Gene mutation in vivo (species)	Other genotoxic end-points in vitro (test system)	Other genotoxic end-points in vivo (species)
N-Methyl-*N*-nitrosourea[186, 187] CAS 684-93-5 (Group 2A)	Mouse, CFW/D, nd, i.p.	Blood, lymphoma, nd Lung, nd Liver, nd Fore-stomach, nd	f, drinking-water Tongue[75] **+f** Pooled oral tissue[75] **+f** (except tongue)	Mouse B6C3F1, m, i.p. Spleen[188, 189] **+m** Lung[188] **+m** Liver[188] **+m** Brain[188] **+m** Germ cells[188] **+m**	Ames test **+** (bacteria) *Hprt* **+** (mamm. cells) Ouabain **+** (mamm. cells) Gene mutation **+** (fungi) Gene mutation **+** (different plants)	Spot test **+** (mouse)[81] SLRL **+** (*Drosophila*) Host-mediated gene mutation **+** (nd)	SCE **+** (mamm. & human cells) Cytogenetic **+** (human & mamm. cells) UDS **+** (human & mamm. cells) Cytogenetic **+** (different plants) Gene conver. **+** (fungi) Cytogenetic **+** (fungi) DNA damage **+** (bacteria) Cell transf. **+** (mamm. cells)	Micronuclei **+** (mouse)[11, 17] SCE **+** (non-human) Dominant lethal **+** (mouse) UDS **+** (mouse) Cytogenetic **+** (*Drosophila*) Heritable translocation **+** (*Drosophila*)
	Mouse, C3HF/Dp, b, i.p.	Thymus, nd Fore-stomach, nd Lung, nd Liver, m Kidney, nd Ovary, f Orbital glands, nd		Mouse, C57BL/6, m, diet Liver[30] **+m** Glandular stomach[30] **(+)m** Fore-stomach[30] **−m** Lung[30] **−m** Kidney[30] **−m**				

Table A1-1 (Contd)

Substance[a]	Carcinogenicity		Transgenic animal models[d]		Genotoxicity			
(IARC classification)	Species,[b] strain, sex, route	Target organs,[c] sex with positive results	Muta™Mouse sex, route, tested organ, result	Big Blue® species, strain, sex, route, tested organ, result	Gene mutation in vitro (test system)	Gene mutation in vivo (species)	Other genotoxic end-points in vitro (test system)	Other genotoxic end-points in vivo (species)
N-Methyl-*N*-nitrosourea *(contd)*	Mouse, Swiss, m, s.c.	Blood, lymphoma, m Connective tissue (local sarcoma), m						
	Mouse, BALB/c, nd, dermal	Skin, nd						

Table A1-1 (Contd)

Substance[a] (IARC classification)	Carcinogenicity		Transgenic animal models[d]		Genotoxicity			
	Species,[b] strain, sex, route	Target organs,[c] sex with positive results	Muta™Mouse sex, route, tested organ, result	Big Blue® species, strain, sex, route, tested organ, result	Gene mutation in vitro (test system)	Gene mutation in vivo (species)	Other genotoxic end-points in vitro (test system)	Other genotoxic end-points in vivo (species)
Mitomycin C[190–192] CAS 50-07-7 (Group 2B)	Mouse, btk, nd, s.c. Rat, CD, b, i.p.	Subcutaneous tissue (local sarcoma), nd Peritoneum (local sarcoma), b	m, i.p. **#** Liver[193] **−m** Bone marrow[193] **−m** (But clastogenic in the same mice in bone marrow)		Ames test **±** (bacteria) Mouse lymphoma **+** (mamm. cells) *Hprt* **±** (mamm. cells) Gene mutation **+** (plant & fungi)	Spot test **+** (mouse)[81, 152] Specific locus test **+** (mouse) Host-mediated gene mutation **+** (bacteria & rodent) SLRL **+** (*Drosophila*)	UDS **+** (human & mamm. cells) Cytogenetic **+** (mamm. cells, plants & fungi) Micronuclei **+** (human cells) SCE **+** (human & mamm. cells) DNA damage **+** (bacteria) Gene conver. **+** (fungi)	Micronuclei **+** (rat[11] & mouse[11, 68]) Cytogenetic **+** (mouse) SCE **+** (human) UDS **+** (mouse) Dominant lethal **+** (rodents) Heritable translocation **+** (mouse)

Table A1-1 (Contd)

Substance[a] (IARC classification)	Carcinogenicity		Transgenic animal models[d]		Genotoxicity			
	Species,[b] strain, sex, route	Target organs,[c] sex with positive results	Muta™Mouse sex, route, tested organ, result	Big Blue® species, strain, sex, route, tested organ, result	Gene mutation in vitro (test system)	Gene mutation in vivo (species)	Other genotoxic end-points in vitro (test system)	Other genotoxic end-points in vivo (species)
4-Nitroquinoline 1-oxide[75, 194, 195] CAS 56-57-5 (no evaluation)	Mouse, dd or Swiss, f, s.c. Mouse, Swiss, b, dermal Rat, 9 strains tested, b, drinking-water	Lung, f Skin, b Tongue, b	f, drinking-water Tongue[75] **+f** Oral cavity[106] **+f** m, i.p. Bone marrow[196] **+m** Lung[196] **+m** Liver[196] **+m** Spleen[196] **−m** Testis[196] **−m** Stomach[196] **−m** Kidney[196] **−m** m, gavage Stomach[196] **+m** Bone marrow[196] **+m** Liver[196] **+m** Lung[196] **+m** Testis[196] **+m**		Ames test **+** (bacteria) *Hprt* **+** (mamm. cells) Gene mutation **+** (fungi & plants)	Spot test **+** (mouse)[106, 152] Host-mediated gene mutation **+** (nd)	Cytogenetic **+** (human & mamm. cells) UDS **+** (human cells) SCE **+** (human & mamm. cells) Cell transf. **+** (mamm. cells) Gene conver. & recombination **+** (fungi) DNA damage **+** (bacteria)	Micronuclei **+** (mouse[11] & rat[28])

Table A1-1 (Contd)

Substance[a] (IARC classification)	Carcinogenicity		Transgenic animal models[d]		Genotoxicity			
	Species,[b] strain, sex, route	Target organs,[c] sex with positive results	Muta™Mouse sex, route, tested organ, result	Big Blue® species, strain, sex, route, tested organ, result	Gene mutation in vitro (test system)	Gene mutation in vivo (species)	Other genotoxic end-points in vitro (test system)	Other genotoxic end-points in vivo (species)
N-Nitrosodiethylamine[197–199] CAS 55-18-5 (Group 2A)	Mouse, nd, nd, gavage	Liver, nd Oesophagus, nd Fore-stomach, nd Lung, nd Blood, lymphoma, nd	f, i.p. Liver[151] **+f** Bone marrow[151] **−f** m, i.p. Liver[150, 200, 201] **+m** Bone marrow[150, 200] **−m** (micronucleus test negative in the same mice[119])		Ames test **+** (bacteria) *Hprt* **+** (mamm. cells) Ouabain **+** (mamm. cells) Mouse lymphoma **+** (mamm. cells) Gene mutation **+** (fungi & plants)	Spot test **+** (mouse)[81] Specific locus test **−** (mouse) SLRL **+** (*Drosophila*) Host-mediated gene mutation **+** (bacteria/nd)	Cytogenetic **+** (mamm. cells) SCE **+** (human & mamm. cells) UDS **+** (mamm. cells) Cell transf. **+** (mamm. cells) DNA damage **+** (bacteria) Gene conver. & recombination **+** (fungi)	Micronuclei **−** (rat[28] & mouse[11, 17]) Dominant lethal **−** (mouse) SCE **+** (non-human) UDS **−** (mouse) Heritable translocation **±** (*Drosophila*)
	Mouse, nd, nd, dermal	Nasal cavity, nd						
	Mouse, B6C3F1, b, i.p.	Liver, b Lung, b						
	Mouse, C57BL, nd, i.p.	Liver, nd						
	Mouse, Swiss, b, s.c.	Lung, b Local sarcoma, nd						

Table A1-1 (Contd)

Substance[a] (IARC classification)	Carcinogenicity		Transgenic animal models[d]		Genotoxicity			
	Species,[b] strain, sex, route	Target organs,[c] sex with positive results	Muta™Mouse sex, route, tested organ, result	Big Blue® species, strain, sex, route, tested organ, result	Gene mutation in vitro (test system)	Gene mutation in vivo (species)	Other genotoxic end-points in vitro (test system)	Other genotoxic end-points in vivo (species)
N-Nitrosodimethylamine[202, 203] CAS 62-75-9 (Group 2A)	Mouse, nd, nd, oral	Vascular system, nd Liver, nd Lung, nd Kidney, nd	nd, inhalation Nasal mucosa[204] + Liver[204] + Lung[204] –	Mouse B6C3F1, f, gavage Liver[57] **+f** Bladder[57] **–f**	Ames test + (bacteria) Forward mutation + (bacteria) *Hprt* + (mamm. cells) Ouabain + (mamm. cells) Mouse lymphoma + (mamm. cells) Gene mutation + (fungi & plants)	Spot test + (mouse)[152] SLRL + (*Drosophila*)	UDS + (human & mamm. cells) Cytogenetic + (mamm. cells) SCE + (human & mamm. cells) Cell transf. ± (mamm. cells) Gene conver./ recombination + (fungi) DNA damage + (bacteria)	Micronuclei + (mouse)[10] Micronuclei ± (rat[28]) UDS – (mouse germ cells) Cytogenetic – (germ cells in mammalia) Dominant lethal ± (rodent) SCE + (non-human) Heritable translocation + (*Drosophila*)
	Mouse, RF, m, drinking-water	Lung, m Vascular system, m	nd, gavage Liver[204] + Nasal mucosa[204] –	m, i.p. Liver[111, 176, 207] **+m**				
	Mouse, DD, nd, s.c.	Vascular system, nd Lung, nd	m, gavage Liver[205] **+m**	C57BL/6, f, diet Liver[4] **+f** Forestomach[4] **–f** Lung[4] **–f**				
	Mouse, Swiss, b, i.p.	Vascular system, b	f, i.p. Liver[206] **+f** Spleen[206] **+f** Kidney[206] **–f** Lung[206] **–f**	m, gavage Liver[208] **+m**				

Table A1-1 (Contd)

Substance[a] (IARC classification)	Carcinogenicity		Transgenic animal models[d]		Genotoxicity			
	Species,[b] strain, sex, route	Target organs,[c] sex with positive results	Muta™Mouse sex, route, tested organ, result	Big Blue® species, strain, sex, route, tested organ, result	Gene mutation in vitro (test system)	Gene mutation in vivo (species)	Other genotoxic end-points in vitro (test system)	Other genotoxic end-points in vivo (species)
N-Nitrosodimethylamine *(contd)*	Rat, nd, nd, oral	Liver, nd Vascular system, nd Kidney, nd Lung, nd		m, i.p. Liver[176, 209, 210] **+m** Kidney[209, 210] **+m** Lung[209, 210] **+m** Bladder[209] **−m** Bone marrow[209] **−m** Testes[209] **−m** Rat F344, m, gavage Liver[211] **+m**				

Table A1-1 (Contd)

Substance[a] (IARC classification)	Carcinogenicity		Transgenic animal models[d]		Genotoxicity			
	Species,[b] strain, sex, route	Target organs,[c] sex with positive results	Muta™Mouse sex, route, tested organ, result	Big Blue® species, strain, sex, route, tested organ, result	Gene mutation in vitro (test system)	Gene mutation in vivo (species)	Other genotoxic end-points in vitro (test system)	Other genotoxic end-points in vivo (species)
N-Nitrosodi-*n*-propylamine[212–215] CAS 621-64-7 (Group 2B)	Mouse, nd, m, s.c. Rat, nd, nd, drinking-water	Nasal cavity, m Intestine, m Liver, m Oesophagus, nd Fore-stomach, nd Liver, nd Nasal cavity, nd Tongue, nd Blood, leukaemia, nd	m, i.p. Liver[216] **+m** Lung[216] **+m** Kidney[216] **+m** Bone marrow[216] **+m** Bladder[216] **−m** Testis[216] **−m**		Ames test **+** (bacteria) *Hprt* **+** (mamm. cells) Ouabain **+** (mamm. cells) Mouse lymphoma **+** (mamm. cells)	nd	Cytogenetic **+** (mamm. cells) UDS **+** (human & mamm. cells) Cell transf. **+** (mamm. cells)	Micronuclei **−** (mouse)[17] SCE **+** (mouse) DNA damage **+** (rat)

Table A1-1 (Contd)

Substance[a] (IARC classification)	Carcinogenicity		Transgenic animal models[d]		Genotoxicity			
	Species,[b] strain, sex, route	Target organs,[c] sex with positive results	Muta™Mouse sex, route, tested organ, result	Big Blue® species, strain, sex, route, tested organ, result	Gene mutation in vitro (test system)	Gene mutation in vivo (species)	Other genotoxic end-points in vitro (test system)	Other genotoxic end-points in vivo (species)
Phenobarbital[217–219] CAS 50-06-6 (Group 2B)	Mouse, CFI, b, diet	Liver, b	m, diet Liver[201, 220] **–m** (but increased liver weight indicating systemic effects)	Mouse C57BL/6, f, diet Liver[6] **–f** Mouse B6C3F1, m, diet Liver[221] **(+)m**	Gene mutation **±** (bacteria) *Hprt* **+** (mamm. cells) Mouse lymphoma **±** (mamm. cells) Other gene mutation **–** (mamm. cells) Gene mutation **+** (human cells) Gene mutation **–** (fungi)	Somatic mutation **–** (*Drosophila*) SLRL **–** (*Drosophila*)	Cytogenetic **+** (human cells) Cytogenetic **±** (mamm. cells) Micronuclei **–** (mamm. cells) SCE **–** (human cells) SCE **±** (mamm. cells) Cell transf. **±** (mamm. cells) UDS **–** (mamm. cells) Gene conver./recombination **–** (fungi) Aneuploidy **+** (fungi) Aneuploidy **–** (mamm. cells)	Micronuclei **±** (mouse)[10, 17] Cytogenetic **–** (mouse) SCE **–** (mouse)
	Rat, Wistar, b, drinking-water	Liver, b						

Table A1-1 (Contd)

Substance[a] (IARC classification)	Carcinogenicity		Transgenic animal models[d]		Genotoxicity			
	Species,[b] strain, sex, route	Target organs,[c] sex with positive results	Muta™Mouse sex, route, tested organ, result	Big Blue® species, strain, sex, route, tested organ, result	Gene mutation in vitro (test system)	Gene mutation in vivo (species)	Other genotoxic end-points in vitro (test system)	Other genotoxic end-points in vivo (species)
Procarbazine[222–224] CAS 671-16-9 (Group 2A)	Mouse, CDF1, b, gavage Mouse, CDF1, m, i.p. Mouse, Swiss, b, i.p.	Lung, b Blood, leukaemia, b Lung, m Blood, leukaemia, m Lung, b Blood, lymphoma, f Kidney, f Uterus, f	m, i.p. Bone marrow[20, 21, 225, 226] **+m** Lung[225] **+m** Spleen[225] **+m** Kidney[225] **+m** Liver[225, 226] **+m** Testis[225] **+m** Brain[225] **−m** (positive micronucleus test at even lower doses[225])		Gene mutation **+** (mamm. cells & fungi) Ames test **±** (bacteria)	Spot test **+** (mouse)[81, 106, 227] Specific locus test **+** (mouse) SLRL **+** (*Drosophila*) Somatic mutation **+** (*Drosophila*) Host-mediated gene mutation **±** (bacteria/ rodent)	Cytogenetic **±** (mamm. cells) Cytogenetic **−** (human cells) SCE **−** (mamm. cells) Cell transf. **−** (mamm. cells) Gene conver./ recombination **+** (fungi) DNA damage **+** (bacteria)	Micronuclei **+** (mouse)[11, 228] Cytogenetic **+** (mouse) SCE **+** (mouse & hamster) Dominant lethal **±** (mouse) Dominant lethal **+** (*Drosophila*) Heritable translocation **−** (mouse & *Drosophila*) DNA damage **+** (rodents)

Table A1-1 (Contd)

Substance[a] (IARC classification)	Carcinogenicity		Transgenic animal models[d]		Genotoxicity			
	Species,[b] strain, sex, route	Target organs,[c] sex with positive results	Muta™Mouse sex, route, tested organ, result	Big Blue® species, strain, sex, route, tested organ, result	Gene mutation in vitro (test system)	Gene mutation in vivo (species)	Other genotoxic endpoints in vitro (test system)	Other genotoxic endpoints in vivo (species)
β-Propiolactone (3-propanolide)[229–232] CAS 57-57-8 (Group 2B) alkylating substance (local effects)	Mouse, nd, nd, i.p. Mouse, nd, nd, dermal Mouse, nd, nd, s.c. Rat, nd, nd, gavage	Liver, nd Blood, lymphoma, nd Skin, nd Local sarcoma, nd Stomach, nd	m, gavage Stomach[182] **+m** Liver[182] **+m** Bone marrow[182] **−m**		Ames test **+** (bacteria) *Hprt* **+** (mamm. cells) Mouse lymphoma **+** (mamm. cells) Gene mutation **+** (fungi)	Host-mediated **+** (bacteria & mouse) SLRL **+** (*Drosophila*) Gene mutation **+** (plant)	Cytogenetic **+** (mamm. cells) UDS **+** (human & mamm. cells) SCE **+** (mamm. cells) DNA damage **+** (bacteria) Cell transf. **+** (mamm. cells) Gene conver./ recombination **+** (fungi)	Micronuclei **−** (mouse)[11, 17] Cytogenetic **+** (plant) Heritable translocation **+** (*Drosophila*) more local than systemic effects

Table A1-1 (Contd)

Substance[a] (IARC classification)	Carcinogenicity		Transgenic animal models[d]		Genotoxicity			
	Species,[b] strain, sex, route	Target organs,[c] sex with positive results	Muta™Mouse sex, route, tested organ, result	Big Blue® species, strain, sex, route, tested organ, result	Gene mutation in vitro (test system)	Gene mutation in vivo (species)	Other genotoxic end-points in vitro (test system)	Other genotoxic end-points in vivo (species)
N-Propyl-*N*-nitrosourea[233, 234] CAS 816-57-9 (no evaluation)	No data available on mice Rat, BUF/MNA, b, drinking-water	No data available on mice Thymus, lymphoma, b Duodenum, m	m, i.p. Bone marrow[235] **+m** Spleen[235] **+m** Liver[235] **+m** Kidney[235] **+m** Lung[235] **+m** Heart[235] **+m** Testis[235] **+m** Brain[235] **−m**		Ames test **+** (bacteria) *Hprt* **+** (mamm. cells)	Spot test **+** (mouse)[81]	UDS **−** (human cells) Gene conver. & recombination **+** (fungi)	nd

Table A1-1 (Contd)

Substance[a] (IARC classification)	Carcinogenicity: Species,[b] strain, sex, route	Carcinogenicity: Target organs,[c] sex with positive results	Transgenic animal models[d]: Muta™Mouse sex, route, tested organ, result	Transgenic animal models[d]: Big Blue® species, strain, sex, route, tested organ, result	Genotoxicity: Gene mutation in vitro (test system)	Genotoxicity: Gene mutation in vivo (species)	Genotoxicity: Other genotoxic endpoints in vitro (test system)	Genotoxicity: Other genotoxic endpoints in vivo (species)
Quinoline[236–240] CAS 91-22-5 (Group 2A)	Mouse, ddy, b, diet Mouse, CD-1, b, i.p. Rat, Wistar, m, diet	Liver, b Liver, m Liver, m	m, i.p. Liver[241] **+m** Testis[241] **–m** Bone marrow[241] **–m** f, i.p. Liver[206] **+f** Kidney[206] **–f** Lung[206] **–f** Spleen[206] **–f**		Ames test **+** (bacteria)	nd	Cytogenetic **+** (mamm. cells) UDS **+** (mamm. cells) SCE **+** (mamm. cells)	Micronuclei **+** (mouse)[242] Micronuclei **–** (rat) Cytogenetic **+** (rat) Cytogenetic **–** (mouse) SCE **+** (rat) SCE **–** (mouse) UDS **(+)** (rat)

Table A1-1 (Contd)

Substance[a] (IARC classification)	Carcinogenicity		Transgenic animal models[d]		Genotoxicity			
	Species,[b] strain, sex, route	Target organs,[c] sex with positive results	Muta™Mouse sex, route, tested organ, result	Big Blue® species, strain, sex, route, tested organ, result	Gene mutation in vitro (test system)	Gene mutation in vivo (species)	Other genotoxic endpoints in vitro (test system)	Other genotoxic endpoints in vivo (species)
Sodium saccharin[243–245] CAS 128-44-9 (Group 3) non-genotoxic mechanisms in carcinogenesis	Rat, Wistar, b, diet	Bladder, m		Rat, F344, m, diet **#** Liver[170] **−m** Bladder[170] **−m**	Ames test **−** (bacteria) Mouse lymphoma **−** (mamm. cells) Gene mutation **+** (human cells) Gene mutation **+** (fungi) Genotoxic effects presumably attributable to impurities[246] or increased osmolarity (in vitro)[243–245]	Spot test **±**[81] (mouse; unknown purity, no dose–response) Somatic mutation **±** (mouse) Host-mediated gene mutation **+** (bacteria & mouse) SLRL **±** (*Drosophila*)	Cytogenetic **+** (human & mamm. cells) SCE **+** (mamm. cells) SCE **±** (human cells) UDS **−** (mamm. cells) Cell transf. **−** (human & mamm. cells) Gene conver./ recombination **±** (fungi) Aneuploidy **+** (fungi)	Micronuclei **−** (mouse)[246] Cytogenetic **−** (hamster & mouse) SCE **+** (hamster) SCE **−** (mouse) Dominant lethal **±** (mouse) Heritable translocation **−** (mouse & *Drosophila*) DNA damage **+** (mouse) DNA damage **−** (rat) DNA binding **−** (rat)
	Rat, CD, b, diet	Bladder, m						
	Mouse, Swiss, f, bladder insertion	Bladder, f						

Table A1-1 (Contd)

Substance[a] (IARC classification)	Carcinogenicity		Transgenic animal models[d]		Genotoxicity			
	Species,[b] strain, sex, route	Target organs,[c] sex with positive results	Muta™Mouse sex, route, tested organ, result	Big Blue® species, strain, sex, route, tested organ, result	Gene mutation in vitro (test system)	Gene mutation in vivo (species)	Other genotoxic end-points in vitro (test system)	Other genotoxic end-points in vivo (species)
Tamoxifen [247–249] CAS 10540-29-1 (Group 1)	Rat, Wistar, b, gavage Rat, Wistar, f, nd	Liver, b Uterus, f		Rat, F344, f, gavage Liver[250] **+f** Rat, F344, f, i.p. Liver[248] **+f** Uterus[248] **–f**	Ames test **–** (bacteria) *Hprt* **–** (mamm. cells)		Micronuclei **+** (human cells) UDS **–** (mamm. cells) Cell transf. **+** (mamm. cells) DNA binding **+** (human & mamm. cells)	Micronuclei **+** (mouse)[251] Cytogenetic **+** (rat & mouse) Aneuploidy **+** (rat) DNA binding **+** (rat & mouse)

Table A1-1 (Contd)

Substance[a] (IARC classification)	Carcinogenicity		Transgenic animal models[d]		Genotoxicity			
	Species,[b] strain, sex, route	Target organs,[c] sex with positive results	Muta™Mouse sex, route, tested organ, result	Big Blue® species, strain, sex, route, tested organ, result	Gene mutation in vitro (test system)	Gene mutation in vivo (species)	Other genotoxic end-points in vitro (test system)	Other genotoxic end-points in vivo (species)
2,3,7,8-Tetrachloro-dibenzo-*p*-dioxin (TCDD)[252] CAS 1746-01-6 (Group 1) "not directly genotoxic" MAK K4	Rat, Sprague-Dawley, b, diet	Liver, f Hard palate, b Nasal turbinates, b Tongue, b Lung, f		Rat F344, b, gavage **#** Liver[253] **–b** (DNA analysis also negative, but increased liver weight indicating hepatic effects)	Ames test **–** (bacteria) Mouse lymphoma **±** (mamm. cells) Ouabain/AraC **–** (mamm. cells)	Spot test **–** (mouse, but promoter activity)[254]	Cell transf. **–** (mamm. cells) Cell transf. **±** (human cells) SCE **+** (human cells) UDS **–** (human cells) Micronucleus **+** (human cells)	Micronuclei **–** (mouse)[255] Cytogenetic **–** (human & mouse) SCE **–** (human, rat & mouse) DNA damage **+** (rat) DNA binding **–** (mouse)
	Rat, Osborne, b, gavage	Thyroid, b Liver, f						
	Mouse, B6C3F1, b, gavage	Liver, b Lung , m Thyroid, f Blood, lymphoma, f Skin, fibro-sarcoma, f						

Table A1-1 (Contd)

Substance[a] (IARC classification)	Carcinogenicity		Transgenic animal models[d]		Genotoxicity			
	Species,[b] strain, sex, route	Target organs,[c] sex with positive results	Muta™Mouse sex, route, tested organ, result	Big Blue® species, strain, sex, route, tested organ, result	Gene mutation in vitro (test system)	Gene mutation in vivo (species)	Other genotoxic endpoints in vitro (test system)	Other genotoxic endpoints in vivo (species)
Tetrachloromethane (carbon tetrachloride)[256] CAS 56-23-5 (Group 2B) MAK K4 (compensatory cell regeneration)	Mouse, B6C3F1, b, gavage *Mouse*, BDF1, b, inhalation Rat, F344, b, inhala*tion*	Liver, b Adrenal gland, b Liver, b Adrenal gland, b Liver, b	m, gavage **#** Liver[116, 220] **−m** (but some regeneration of *the liver* detected in histopathology[220])		Ames test **−** (bacteria)	SLRL **−** (*Drosophila*)	Micronuclei **±** (human cells) Aneuploidy **±** (mamm. cells) DNA damage **(+)** (mamm. cells) DNA damage **−** (bacteria) Mitotic *recombination* **+** (fungi)	Micronuclei **−** (mouse)[17] Cytogenetic **−** (rat & mouse) SCE **−** (*rat* & mouse) DNA damage **−** (rat) DNA adducts **±** (rodents)

Table A1-1 (Contd)

Substance[a] (IARC classification)	Carcinogenicity		Transgenic animal models[d]		Genotoxicity			
	Species,[b] strain, sex, route	Target organs,[c] sex with positive results	Muta™Mouse sex, route, tested organ, result	Big Blue® species, strain, sex, route, tested organ, result	Gene mutation in vitro (test system)	Gene mutation in vivo (species)	Other genotoxic end-points in vitro (test system)	Other genotoxic end-points in vivo (species)
Trichloroethylene[257] CAS 79-01-6 (Group 2A)	Mouse, NMRI, b, inhalation Mouse, ICR, f, inhalation Mouse, Swiss, b, inhalation Mouse, B6C3F1, b, inhalation Mouse, B6C3F1, b, gavage	Blood, lymphoma, f Lung, f (but stabilized with epichlorohydrin) Liver, m Lung, m Liver, f Lung, f Liver, b	b, inhalation **#** Bone marrow[258] **–b** Kidney[258] **–b** Spleen[258] **–b** Liver[258] **–b** Lung[258] **–b** m, inhalation Testis[258] **–m** MTD possibly not reached		Ames test **(+)** (bacteria) Mouse lymphoma **±** (mamm. cells) Gene mutation **(+)** (fungi)	Spot test **+** (mouse,[81] possibly contaminated with epoxides[257]) Host-mediated gene mutation **–** (nd)	Cytogenetic **–** (mamm. cells) UDS **–** (mamm. cells) Cell transf. **(+)** (mamm. cells) Aneuploidy **+** (fungi)	Micronuclei **+** (mouse)[10, 124] Micronuclei **±** (rat) Cytogenetic **–** (mouse & rat) SCE **–** (rat & mouse) Dominant lethal **–** (mouse) UDS **–** (mouse) DNA damage **+** (mouse & rat)

Table A1-1 (Contd)

Substance[a]	Carcinogenicity		Transgenic animal models[d]		Genotoxicity			
(IARC classification)	Species,[b] strain, sex, route	Target organs,[c] sex with positive results	Muta™Mouse sex, route, tested organ, result	Big Blue® species, strain, sex, route, tested organ, result	Gene mutation in vitro (test system)	Gene mutation in vivo (species)	Other genotoxic endpoints in vitro (test system)	Other genotoxic endpoints in vivo (species)
Tris(2,3-dibromopropyl)phosphate[259–261] CAS 126-72-7 (Group 2A)	Mouse, B6C3F1, b, diet Rat, F344, b, diet	Lung, b Kidney, m Forestomach, b Liver, f Kidney, b		Mouse, B6C3F1, m, gavage Kidney[65, 262] **+m** Liver[65, 262] **−m** Stomach[65, 262] **−m**	Ames test **+** (bacteria) Mouse lymphoma **+** (mamm. cells) *Hprt* **+** (mamm. cells)	SLRL **+** (*Drosophila*)	Cytogenetic **+** (mamm. cells) Cytogenetic – (human cells) SCE **+** (human & mamm. cells) DNA damage **+** (human & mamm. cells) UDS **+** (mamm. cells) Cell transf. **±** (mamm. cells)	Micronuclei **−** (mouse)[17, 259–261] Micronuclei **+** (hamster) Cytogenetic **−** (mouse & rat) DNA damage **+** (rat) Heritable translocation **+** (*Drosophila*) Mitotic recombination **+** (*Drosophila*)

Table A1-1 (Contd)

Substance[a] (IARC classification)	Carcinogenicity		Transgenic animal models[d]		Genotoxicity			
	Species,[b] strain, sex, route	Target organs,[c] sex with positive results	Muta™Mouse sex, route, tested organ, result	Big Blue® species, strain, sex, route, tested organ, result	Gene mutation in vitro (test system)	Gene mutation in vivo (species)	Other genotoxic end-points in vitro (test system)	Other genotoxic end-points in vivo (species)
Urethane (ethyl carbamate)[263–265] CAS 51-79-6 (Group 2B)	Mouse, Swiss, b, drinking-water	Blood, lymphoma & leukaemia, b Liver, haemangioma, b Skin, b	i.p., m Lung[266] **+m** Liver[266] **+m** Spleen[266] **+m** Bone marrow[266] **+m**	Mouse, C57BL/6, f, diet Lung[30] **+f** Liver[30] **+f** Forestomach[4] **+f**	Ames test **–** (bacteria) Reverse mutation **–** (*E. coli*, bacteria) Gene mutation **+** (fungi)	Host-mediated gene mutation **–** (bacteria/nd) SLRL **+** (*Drosophila*)	Cytogenetic **+** (mamm. cells) SCE **+** (human & mamm. cells) UDS **±** (mamm. cells) Cell transf. **+** (mamm. cells) Gene conver./ recombination **±** (fungi) DNA damage **–** (bacteria)	Micronuclei **+** (rat & mouse)[10] SCE **+** (non-human) Cytogenetic **±** (*Drosophila*) Heritable translocation **+** (*Drosophila*)
	Mouse, CTM, b, drinking-water	Lung, m Liver, m						
	Mouse, C3H, m, i.p.	Lung, b Thymus, b Liver, nd Harderian gland, nd						

Table A1-1 (Contd)

1: Shephard et al. (1993); **2**: HSDB (2003b); **3**: NTP (2003a); **4**: Brooks et al. (1995); **5**: Ross & Leavitt (1998); **6**: Gunz et al. (1993); **7**: Hüttner et al. (1988); **8**: Fahrig (1988); **9**: Gocke et al. (1983); **10**: Mavournin et al. (1990); **11**: Heddle et al. (1983); **12**: Schinz et al. (1955); **13**: Morris et al. (1960); **14**: GENE-TOX (1998a); **15**: CCRIS (1995a); **16**: Tinwell et al. (1998); **17**: Morita et al. (1997a, 1997b); **18**: IARC (1994a); **19**: CCRIS (1996); **20**: Myhr (1991); **21**: Hoorn et al. (1993); **22**: Krebs & Favor (1997); **23**: Neuhäuser-Klaus & Schmahl (1989); **24**: IARC (1993a); **25**: Autrup et al. (1996); **26**: Dycaico et al. (1996); **27**: Davies et al. (1997); **28**: Trzos et al. (1978); **29**: IARC (1983); **30**: Shephard et al. (1995); **31**: IARC (1971); **32**: GENE-TOX (1998b); **33**: CCRIS (2003a); **34**: Fletcher et al. (1998); **35**: IARC (1993c); **36**: Nagao et al. (1998); **37**: Suzuki et al. (1996b); **38**: IARC (1993e); **39**: Tucker et al. (1989); **40**: Loprieno et al. (1991); **41**: Hoshi et al. (2004); **42**: Itoh et al. (2000); **43**: IARC (1993d); **44**: Nagao (1999); **45**: Lynch et al. (1998); **46**: Okonogi et al. (1997); **47**: Stuart et al. (2001); **48**: Okochi et al. (1999); **49**: Yang et al. (2002); **50**: Stuart et al. (2000b); **51**: Zhang et al. (1996); **52**: IARC (1993b); **53**: Davis et al. (1996); **54**: Bol et al. (2000); **55**: Wild et al. (1985); **56**: European Chemicals Bureau (2002); **57**: Ashby et al. (1994); **58**: IARC (1987a); **59**: IARC (1987b); **60**: IARC (1977a); **61**: Rihn et al. (2000); **62**: Topinka et al. (2004c); **63**: IARC (1982); **64**: IPCS (1993); **65**: Provost et al. (1996); **66**: Mullin et al. (1998); **67**: Mullin et al. (1995); **68**: Shelby & Witt (1995); **69**: IPCS (1998b); **70**: CCRIS (2003b); **71**: Hakura et al. (1998); **72**: Hakura et al. (1999); **73**: Mientjes et al. (1996); **74**: Kosinska et al. (1999); **75**: von Pressentin et al. (1999); **76**: Skopek et al. (1996); **77**: Kohler et al. (1991a); **78**: Kohler et al. (1991b); **79**: Shane et al. (1997); **80**: Shane et al. (2000a); **81**: Styles & Penman (1985); **82**: Shimada et al. (1992); **83**: IRIS (2002a); **84**: CCRIS (2001a); **85**: IPCS (1995a); **86**: Pletsa et al. (1999); **87**: Araki et al. (1995); **88**: IARC (1999a); **89**: Recio et al. (1993); **90**: Sisk et al. (1994); **91**: Recio et al. (1996); **92**: Adler et al. (1994); **93**: Cunningham et al. (1986); **94**: IARC (1981a); **95**: IARC (1987c); **96**: GENE-TOX (1998c); **97**: CCRIS (1995b); **98**: MAK (2000); **99**: Butterworth et al. (1998); **100**: Robbiano et al. (1998); **101**: IARC (1981b); **102**: IARC (1987d); **103**: Gorelick et al. (1999); **104**: Walker et al. (1999a); **105**: Hoyes et al. (1998); **106**: Hart (1985); **107**: Machemer & Lorke (1978); **108**: MAK (1994); **109**: Loveday et al. (1990); **110**: Suter et al. (1996); **111**: Cunningham et al. (1996); **112**: Hayward et al. (1995); **113**: NTP (2003b); **114**: Shelby et al. (1993); **115**: IARC (1999d); **116**: Hachiya & Motohashi (2000); **117**: Schmezer et al. (1998a); **118**: Asita et al. (1992); **119**: IARC (1999e); **120**: Sasaki et al. (1986); **121**: Albanese et al. (1988); **122**: Belitsky et al. (1994); **123**: IARC (1999f); **124**: IARC (1979); **125**: MAK (2002a); **126**: Fahrig & Steinkamp-Zucht (1996); **127**: Fletcher et al. (1999); **128**: Lefevre et al. (1997); **129**: CCRIS (2002a); **130**: GENE-TOX (1998d); **131**: Ashby et al. (1993); **132**: Hachiya et al. (1999); **133**: Suzuki et al. (1999a); **134**: Gorelick et al. (1995); **135**: Lonardo et al. (1996); **136**: Thompson & Osterhues (1995); **137**: Shelton et al. (2000); **138**: Manjanatha et al. (1998); **139**: Manjanatha et al. (2000); **140**: IARC (1994b); **141**: Sisk et al. (1997); **142**: Recio et al. (2004); **143**: Walker et al. (1997); **144**: Jenssen & Ramel (1980); **145**: IARC (1974a); **146**: DECOS (1989); **147**: CCRIS (2001b); **148**: GENE-TOX (1995a); **149**: Suzuki et al. (1997); **150**: Suzuki et al. (1994); **151**: Mientjes et al. (1998); **152**: Fahrig (1977); **153**: IARC (1978a); **154**: GENE-TOX (1998e); **155**: JEMS/MMS (1996); **156**: Douglas et al. (1995a); **157**: Gorelick et al. (1997); **158**: Putman et al. (1997); **159**: Winegar et al. (1997); **160**: Katoh et al. (1997); **161**: Zimmer et al. (1999); **162**: Skopek et al. (1995); **163**: Provost & Short (1994); **164**: Shibuya & Morimoto (1993); **165**: IARC (2001); **166**: BUA (1996); **167**:

Douglas et al. (1995b); **168**: Neuhäuser-Klaus & Chauhan (1987); **169**: IPCS (1998a); **170**: Turner et al. (2001); **171**: IARC (1999b); **172**: IARC (1974b); **173**: Renault et al. (1997); **174**: Itoh et al. (1997); **175**: Brooks & Dean (1997); **176**: Mirsalis et al. (1993); **177**: Jenssen & Ramel (1976); **178**: IARC (1974c); **179**: IARC (1987e); **180**: IARC (1987f); **181**: Brooks & Dean (1996); **182**: Brault et al. (1996); **183**: IARC (1985); **184**: CCRIS (2003f); **185**: Hashimoto et al. (2004); **186**: IARC (1978b); **187**: GENE-TOX (1998f); **188**: Provost et al. (1993); **189**: Monroe et al. (1998); **190**: IARC (1976); **191**: CCRIS (2002b); **192**: GENE-TOX (1998g); **193**: Suzuki et al. (1993); **194**: GENE-TOX (1998h); **195**: CCRIS (2003c); **196**: Nakajima et al. (1999); **197**: IARC (1978c); **198**: CCRIS (2003d); **199**: GENE-TOX (1998i); **200**: Suzuki et al. (1995); **201**: Okada et al. (1997); **202**: IARC (1978d); **203**: GENE-TOX (1995b); **204**: Schmezer et al. (1998b); **205**: Tinwell et al. (1994a); **206**: Suzuki et al. (1998); **207**: Shane et al. (2000b); **208**: Tinwell et al. (1994b); **209**: Suzuki et al. (1996a); **210**: Wang et al. (1998); **211**: Gollapudi et al. (1998); **212**: MAK (1991); **213**: CCRIS (2002c); **214**: GENE-TOX, (1998j); **215**: IRIS (2002b); **216**: Itoh et al. (1999); **217**: IARC (1977b); **218**: IARC (1987g); **219**: IARC (1987h); **220**: Tombolan et al. (1999a); **221**: Shane et al. (2000c); **222**: IARC (1981c); **223**: IARC (1987i); **224**: Tweats & Gatehouse (1988); **225**: Suzuki et al. (1999b); **226**: Pletsa et al. (1997); **227**: Neuhäuser (1977); **228**: Kliesch et al. (1981); **229**: MAK (1976); **230**: ACGIH (1998); **231**: CCRIS (1995c); **232**: GENE-TOX (1995c); **233**: CCRIS (2000); **234**: GENE-TOX (1992); **235**: Hara et al. (1999); **236**: IRIS (2002c); **237**: CCRIS (2003e); **238**: GENE-TOX (1995d); **239**: HSDB (2003a); **240**: Matsuoka et al. (1979); **241**: Miyata et al. (1998); **242**: Hamoud et al. (1989); **243**: IARC (1980); **244**: IARC (1987j); **245**: IARC (1999c); **246**: Eckhardt et al. (1980); **247**: IARC (1996); **248**: da Costa et al. (2002); **249**: CCRIS (2001d); **250**: Davies et al. (1997); **251**: Vijayalaxmi & Rai (1996); **252**: IARC (1997); **253**: Thornton et al. (2001); **254**: Fahrig (1993); **255**: Meyne et al. (1985); **256**: MAK (2002b); **257**: MAK (1998); **258**: Douglas et al. (1999); **259**: IPCS (1995b); **260**: CCRIS (2001c); **261**: GENE-TOX (1998k); **262**: de Boer et al. (1996); **263**: IARC (1974d); **264**: GENE-TOX (1998l); **265**: Ishidate & Odashima (1977); **266**: Williams et al. (1998)

APPENDIX 2: THE *CII* ASSAY IN TRANSGENIC RODENT STUDIES

The *λcII* assay was introduced in section 5.1. It is available in the Muta™Mouse and in Big Blue® mice and rats. The data pool on this test system is growing fast, because this transgene test system is now preferred in Muta™Mouse and Big Blue® mice and rats instead of the *lacZ* and *lacI* transgene. In Table A2-1, examples of studies are given.

Table A2-1. Examples of studies using the *cII* transgene in vitro or in vivo in either Muta™Mouse or Big Blue® rodents for investigation of mutagenic activity

Chemical	Muta™Mouse tested organ	Result	Big Blue® species / tested organ	Result	Remarks	Reference
None (spontaneous mutations)			Mouse Liver Lung Spleen		Type of mutations studied in SA	Harbach et al. (1999)
Acrylamide and the epoxide metabolite glycidamide			Mouse Embryonic fibroblasts	+	In vitro study; other parameters tested parallel: MS & DNA adducts; the effects of glycidamide were more pronounced	Besaratinia & Pfeifer (2003, 2004)
o-Aminoazotoluene	Liver Colon	+ +			SA performed, G:C→T:A transversions induced	Kohara et al. (2001)
2-Amino-3-methyl-imidazo[4,5-*f*]quino-line (IQ)			Rat Liver Colon	 + +	Other parameters tested parallel: MS, DNA strand breaks & adducts, and oxidative stress	Hansen et al. (2004)
2-Amino-3-methyl-imidazo[4,5-*f*]quin-oline (IQ)			Rat Colon Liver	 + +	Other parameters tested parallel: MS, DNA adducts and strand breaks	Møller et al. (2002)
2-Amino-1-methyl-6-phenylimidazo[4,5-*b*]pyridine (PhIP)	Small intestine	+			Comparison with *lacZ*; also other heterocyclic amines tested; SA performed	Itoh et al. (2003)

Table A2-1 (Contd)

Chemical	Muta™Mouse tested organ	Result	Big Blue® species / tested organ	Result	Remarks	Reference
2-Amino-1-methyl-6-phenylimidazo[4,5-*b*]pyridine (PhIP)			Rat Colon	 +	SA performed; MS not different in *lacI* and *cII* gene in this model for intergenic mutational analysis	Stuart et al. (2000a)
Aristolochic acid	Forestomach Kidney Bladder Colon Glandular stomach Lung Liver Bone marrow Spleen Testis	+ + + + (+) (+) (+) (+) (+) –			Micronucleus induction in peripheral blood of the same mice studied (–); comparison with *lacZ*; SA performed, G:C→T:A transversions induced	Kohara et al. (2002a)
Benzo[*f*]quinoline	Liver Spleen Lung Kidney Bone marrow	– – – – –			Parallel measurement of mutagenic activity in *lacZ* (also negative results); SA performed	Yamada et al. (2004)

Table A2-1 (Contd)

Chemical	Muta™Mouse tested organ	Result	Big Blue® species / tested organ	Result	Remarks	Reference
Benzo[*h*]quinoline	Liver Spleen Lung Kidney Bone marrow	– – – – –			Parallel measurement of mutagenic activity in *lacZ* (positive results in the lung, other organs negative); SA performed	Yamada et al. (2004)
Bitumen fumes			Mouse Lung	 –	Also screening of DNA adducts in the lung (also negative)	Micillino et al. (2002)
Diesel exhaust particles	Lung	–			DNA strand breaks (comet assay) due to inflammation detected in the lung of mice but no mutation	Dybdahl et al. (2004)
Diesel exhaust particles			Rat Lung	 –	Oral application; no mutations, although DNA strand breaks (comet assay) and DNA adducts were detected in the lung	Müller et al. (2004)
Dimethylnitrosamine			Mouse Liver	 +	SA revealed differences between MS in *lacI* and *cII*	Shane et al. (2000b)
Dinitropyrene mixture	Stomach Colon Bone marrow Lung	+ + + +			Micronucleus induction in peripheral blood of the same mice studied (−); comparison with *lacZ*; SA performed,	Kohara et al. (2002b)

Table A2-1 (Contd)

Chemical	Muta™Mouse tested organ	Result	Big Blue® species / tested organ	Result	Remarks	Reference
(contd)	Liver	–			G:C→T:A transversions induced	
N-Ethyl-*N*-nitroso-urea			Mouse Liver Spleen Lung	 + + +	Comparison of mutant frequency in *lacI* and *cII*; similar results, although the fold increase over control was higher in *lacI*	Zimmer et al. (1999)
N-Ethyl-*N*-nitroso-urea			Mouse Brain	 +	Mutation induced in prenatal and 8 days postnatal exposed mice but not in adults; SA performed	Slikker et al. (2004)
N-Ethyl-*N*-nitroso-urea			Mouse & rat Embryonic fibroblasts	 +	In vitro models for molecular toxicology studies	Erexson et al. (1999)
N-Ethyl-*N*-nitroso-urea			Mouse Intestine	 +	Big Blue® mice crossed with Bloom Syndrome mice; F2 used plus wild-type; parallel study of micronuclei and loss of heterozygosity	Wang & Heddle (2004)
N-Ethyl-*N*-nitroso-urea			Mouse Liver Spleen Bone marrow	 + + +	Time course of manifestation studied in different organs; tissue-specific effects	Wang et al. (2004)

Table A2-1 (Contd)

Chemical	Muta™Mouse tested organ	Result	Big Blue® species / tested organ	Result	Remarks	Reference
α-Hydroxytamoxifen			Rat Liver	 +	Tamoxifen itself was less mutagenic than this metabolite; comparison with *lacI*; SA performed	Chen et al. (2002)
Leucomalachite green			Mouse Liver Rat Liver	 + –	Micronucleus induction in peripheral blood of the same mice negative as well as the *Hprt* assay; no mutagenic effects with malachite green; SA performed	Mittelstaedt et al. (2004)
4-(Methylnitros-amino)-1-(3-pyridyl)-1-butanone	Lung Liver	+ +			Also *lacZ* studied; SA performed; predominantly A:T→T:A and/or A:T→C:G mutations	Hashimoto et al. (2004)
3-Nitrobenzanthrone	Colon Liver Bladder Lung Kidney Spleen Testis	+ + + – – – –			Other parameters tested in parallel: MS, DNA adducts, micronuclei in peripheral blood	Arlt et al. (2004)

Table A2-1 (Contd)

Chemical	Muta™Mouse tested organ	Result	Big Blue® species / tested organ	Result	Remarks	Reference
6-Nitrochrysene			Rat Mammary	 +	Other parameters tested in parallel: MS, DNA adducts, mammary tumour formation	Boyiri et al. (2004)
1,7-Phenanthroline	Liver Lung Spleen Kidney Bone marrow	+ − − − −			Parallel measurement of mutagenic activity in *lacZ* (positive results in liver & lung); SA performed: G:C→C:G transversions	Yamada et al. (2004)
Polyphenon E (green tea catechin mixture)			Mouse Liver Lung Spleen	 − − −	Significant increase in mutations in the mouse lymphoma assay in vitro but negative results in Big Blue®	Chang et al. (2003)
Tamoxifen			Rats Liver	 +	Differences in SA detected between *lacI* and *cII* gene	Davies et al. (1999)
Ultraviolet A radiation			Mouse Embryonic fibroblasts	 +	In vitro study; other parameters tested parallel: MS & DNA adducts	Besaratinia et al. (2004)

SA: DNA sequence analysis; MS: mutation spectrum (studied in SA); +: positive (increase in mutant frequency or mutation frequency); −: negative; (+): weak positive

RESUME

Ce document a pour but d'initier aux tests de mutagénicité transgéniques ceux pour qui ce domaine est nouveau et de déterminer quel rôle ces tests pourraient jouer dans les études toxicologiques et la recherche mécanistique.

Un animal transgénique est porteur, dans toutes ses cellules, d'un ADN étranger intégré à l'ADN de ses chromosomes. Dans les tests de mutagénicité transgéniques, l'ADN étranger consiste en un gène étranger (transgène) que l'on injecte dans le noyau d'un embryon de rongeur fécondé. Ces gènes, appelés « gènes rapporteurs » sont transmis par les cellules germinales; ils sont donc présents dans toutes les cellules du rongeur nouveau-né et peuvent servir à détecter les mutations et évaluer leur fréquence.

La première partie du document (chapitres 2 à 6) consiste en un bref survol des tests de génotoxicité in vivo. On y explique comment sont élaborés ces animaux transgéniques en donnant des détails sur les constructions d'ADN utilisées et leur insertion dans les cellules de l'animal receveur. Un certain nombre de modèles transgéniques sont décrits à titre d'exemple – notamment le modèle *lacI*, commercialisé sous les noms de test Big Blue® sur souris ou rat, le modèle *lacZ*, commercialisé sous le nom de Muta™Mouse – ainsi que d'autres modèles de développement plus récent tels que le λ*cII*, le *gpt* delta, le plasmide *lacZ* et le ΦX174.

La capacité d'une telle étude de mutagénicité à déterminer valablement si le composé à expertiser est positif ou négatif est très dépendante de sa conception. Le choix du gène, de l'espèce et du tissu qui seront la cible de l'action mutagène de la substance à expertiser doit reposer sur la connaissance préalable que l'on peut avoir des paramètres pharmacologiques et toxicologiques de cette substance. La dose, la posologie et la durée d'échantillonnage après traitement qui permettent une détermination optimale de la fréquence mutationnelle varient selon la nature du tissu et de l'agent à expertiser, aussi a t-on recommandé un protocole expérimental qui optimalise la détection de tous les agents mutagènes, quelle que soit l'intensité de leur action ou la nature du tissu où elle s'exerce. Un résultat négatif obtenu en appliquant un protocole expérimental robuste doit être considéré comme valable.

La deuxième partie (chapitres 7 à 10) fait la synthèse des résultats publiés au sujet des produits chimiques testés au moyen des modèles *lacI* et *lacZ*, compare ces données avec celles qui ont été obtenues à l'aide des systèmes d'épreuve classiques et en analyse les conséquences. Le choix de ces modèles tient au fait qu'ils sont les seuls pour lesquels on dispose de données suffisantes pour pouvoir procéder à des comparaisons et à des analyses.

A la lumière des données limitées dont on dispose, il semble qu'il y ait un bon accord entre ces résultats et ceux que donnent les tests Muta™Mouse et le test sur souris (ou rat) Big Blue®. Toute divergence observée entre le test Muta™Mouse et le test Big Blue® sur souris peut vraisemblablement être attribuée à des différences dans le protocole expérimental utilisé pour les études en cause plutôt qu'à une différence dans la sensibilité intrinsèque des transgènes rapporteurs.

On a comparé, pour 44 substances, les résultats des tests de mutagénicité transgéniques à ceux du test des micronoyaux sur moëlle osseuse de souris. Si, dans la plupart des cas, les résultats étaient similaires du fait que beaucoup des produits chimiques étudiés étaient fortement cancérogènes, les tests se sont révélés complémentaires – en ce sens que la mise en évidence de la cancérogénicité des produits a été sensiblement meilleure lorsque les deux types de test ont été utilisés. Il semble donc que ce résultat confirme l'avantage théorique qu'il y a à utiliser deux tests qui portent chacun sur un point d'aboutissement différent de l'action génotoxique. La capacité des tests de mutagénicité transgéniques à détecter des mutations géniques dans de multiples tissus constitue également un réel avantage.

Bien que, selon les recommandations de l'Organisation pour la coopération et de développement économiques (OCDE), le spot test sur souris soit l'un des tests de référence en matière de génotoxicité, ce système d'épreuve n'a été que rarement utilisé au cours des dernières décennies pour la détection de mutations somatiques in vivo. La comparaison qui est faite dans ce document entre les deux types de systèmes montre que le test sur souris transgéniques a plusieurs avantages par rapport au spot test et qu'il peut parfaitement le remplacer pour la détection des mutations géniques in vivo, à l'exception toutefois des mutations chromosomiques.

Malgré les différences entre les propriétés mutationnelles des divers modèles d'agents mutagènes utilisés, la réponse des locus exogènes (transgènes *lacI*, *lacZ*) et des locus endogènes (*Dlb-1*, *Hprt*), était en règle générale qualitativement similaire après le traitement. Plusieurs études donnent à penser que la fréquence plus faible des mutants somatiques parmi les gènes endogènes pourrait conférer une meilleure sensibilité dans de telles conditions. Il est toutefois difficile de comparer les transgènes et les gènes endogènes en raison des différences qui existent entre les protocoles expérimentaux optimaux relatifs aux divers types de gènes; dans le cas des transgènes neutres, la durée d'administration plus longue qui est actuellement recommandée augmente la sensibilité de détection.

D'après les données limitées dont on dispose concernant la comparaison des modèles *lacI* et *lacZ* avec le test de synthèse non programmée de l'ADN (test UDS), il semble que les tests sur animaux transgéniques aient une meilleure prédictivité que le test UDS, qui mesure les lésions de l'ADN. Les résultats obtenus avec des animaux transgéniques (*lacI* et *lacZ*) sur plus de 50 substances chimiques concordent avec les données in vitro concernant les mutations géniques, les aberrations chromosomiques et la mesure directe et indirecte des lésions causées à l'ADN par ces composés. L'un des grands avantages des tests de mutagénicité sur souris ou rats transgéniques par rapport aux autres tests de mutagénicité in vivo, c'est que les premiers peuvent détecter les manifestations de la mutagénèse dans n'importe quel organe. On a donc procédé à une analyse afin de déterminer si les études de cancérogénicité au moyen de ces tests de mutagénicité transgéniques permettaient de prédire quels seraient les organes cibles. Dans la plupart des cas, on a trouvé des mutations au niveau des organes cibles de ces études. Pour plusieurs agents cancérogènes présumés génotoxiques, les tests de mutagénicité transgéniques ont révélé la présence de mutations dans des organes qui n'étaient pas les organes cibles des études de cancérogénicité. Comme on l'a observé dans le cas de plusieurs composés, cela ne s'explique probablement pas par une spécificité insuffisante pour les organes cibles de la cancérogénèse. On est plutôt tenté de conclure que la génotoxicité s'exprime au niveau de plusieurs organes mais qu'en raison d'autres facteurs, il n'y a pas apparition de tumeurs dans tous ces organes. Les agents cancérogènes qui ne sont pas présumés génotoxiques donnent généralement des résultats négatifs dans les tests sur animaux transgéniques. On ne possède que très peu de données sur les substances qui donnent des résultats résultats négatifs dans les tests de cancérogénicité sur la souris. Cela étant, ces quelques substances non cancérogènes ont également donné des résultats négatifs dans les tests sur souris transgéniques. Quoi qu'il en soit, il apparaît, à la lumière des données disponibles, que les tests de cancérogénicité transgéniques ont une sensibilité et une prédictivité élevées.

Dans la **partie III** (chapitre 11), sont décrites des études dans lesquelles on utilisé des tests de mutagénicité transgéniques (en particulier les modèles *lacI* et *lacZ* utilisant le gène *cII* et le système murin *gpt* delta) pour des recherches sur le mécanisme de la mutagénèse. En raison de la facilité avec laquelle on peut séquencer le gène *cII* pour l'obtention de spectres mutationnels, il remplace de plus en plus le *lacI* et le *lacZ* dans les tests Muta™Mouse et Big Blue® pour les études de séquençage. Le modèle *gpt* delta est également utilisé en raison de sa facilité de séquençage et notamment aussi, parce qu'il permet de déceler des délétions beaucoup plus importantes qu'avec tous les autres tests, sauf le plasmide *lacZ*.

On a étudié les mutations spontanées avec presque tous les tests de mutagénicité sur animaux transgéniques: *lacZ*, *lacI* et *cII*, plasmide *lacZ* et souris *gpt* delta. Dans tous les systèmes, la mutation spontanée prédominante consiste en transitions G:C→A:T, qui pour la plupart, se produisent au niveau des sites 5'-CpG, ce qui donne à penser que le mécanisme principal de la mutagénèse serait une désamination de la 5-méthylcytosine.

On a étudié la fréquence et la nature des mutations spontanées. Le taux de mutation que l'on en déduit est tributaire de facteurs tels que le site d'insertion du transgène, l'âge, le tissu et la souche. Environ la moitié des mutations se produisent au cours du développement (et la moitié de ces dernières in utero). Plusieurs études ont été consacrées à la nature et à la fréquence des mutations spontanées en fonction de l'âge dans toutes sortes de tissus. Elles ont montré, à l'exception de celles qui utilisaient des souris transgéniques transfectées par un plasmide, que chez les animaux adultes, le spectre des mutations correspondait à l'âge et au type de tissu. Il ne variait pas avec le sexe ou le patrimoine génétique. La fréquence des mutations a toujours été la plus faible dans la lignée germinale mâle et elle est restée pratiquement identique à un âge avancé.

Les tests sur animaux transgéniques se sont révélés intéressants pour l'étude des paradigmes fondamentaux de la toxicologie génétique. Récemment, des études utilisant ces systèmes ont porté sur les points suivants 1) relation dose-réponse des agents cancérogènes génotoxiques et 2) relation entre la formation d'adduits de l'ADN, la fréquence des mutations et les cancers chez les rongeurs. Ces tests sur muridés transgéniques ont également d'autres applications importantes dans la recherche fondamentale sur l'origine des mutations et le rôle préventif à leur égard de divers processus biologiques. Ces travaux concernent l'étude des mécanismes de réparation de l'ADN, la cancérogénèse, le vieillissement et les affections génétiques héréditaires en rapport avec ces processus.

Les spectres mutationnels tirés des données de séquençage de l'ADN ne sont pas jugés indispensables pour l'évaluation des mutations géniques in vivo lorsque le résultat est clairement positif ou négatif, mais ils sont utiles pour l'étude des facteurs liés au mécanisme de la mutagenèse. La possibilité de séquencer les mutations induites au niveau des transgènes rapporteurs permet au chercheur d'obtenir des informations importantes sur plusieurs aspects des mutations. Le document donne quelques exemples de travaux qui montrent comment les tests sur animaux transgéniques et l'analyse ultérieure du spectre mutationnel peuvent être utilisés pour étudier divers aspects de l'activité des agents mutagènes: par exemple 1) la correction clonale et la correction des mutations ex vivo; 2) les lésions prémutagènes; 3) la réponse tissulaire spécifique; 4) l'évaluation des agents génotoxiques qui n'interagissent pas avec l'ADN; 5) la détermination des constituants actifs d'un mélange; 6) la détermination des métabolites actifs; 7) l'étude du mécanismes des mutations par délétion in vivo.

La partie IV (chapitres 12 à 14) porte sur l'utilisation des tests de mutagénicité transgéniques en toxicologie et pour l'évaluation du risque et indique ce qu'ils peuvent apporter de plus dans ces domaines. Ces tests n'ont pas encore été très utilisés par l'industrie pour les contrôles toxicologiques, en grande partie du fait que l'OCDE n'a pas encore élaboré de ligne directrice à cet égard. Il y a peu, un protocole normalisé a été recommandé (Thybaud et al., 2003), qui pourrait servir de base à cette ligne directrice.

Le groupe de travail de l'IPCS/PISC recommande la préparation d'une telle ligne directrice. Son intérêt tient en partie au fait que les tests sur animaux transgéniques sont capables de mettre en évidence les mutations géniques. Si un tel protocole est utilisé, tout résultat négatif pourra être considéré comme fiable.

Le groupe de travail de l'IPCS/PISC recommande également d'inclure les tests de mutagénicité transgéniques dans son dispositif qualitatif pour la mutagénicité et d'autres stratégies de contrôle.

En ce qui concerne les travaux futurs, le groupe de travail de l'IPCS/PISC recommande d'expertiser un certain nombre d'agents dont la non-cancérogénicité est bien établie en utilisant un protocole expérimental robuste (par ex. celui de Thybaud et al., 2003). Il estime également qu'il faudrait recommander d'utiliser les tests de mutagénicité transgéniques pour l'étude des relations mécanistiques entre les mutations et la cancérogenèse et celle de la mutagénèse dans les lignées germinales.

RESUMEN

El presente documento tiene por objeto introducir a los profanos en el campo de las valoraciones de la mutagenicidad transgénica y evaluar su posible función en las pruebas de toxicología y la investigación mecanicista.

Los animales transgénicos tienen ADN extraño, integrado en el de sus cromosomas y presente en todas sus células. En las valoraciones de la mutagenicidad transgénica, el ADN extraño es un gen exógeno (transgén) inyectado en el núcleo de un embrión de roedor fecundado. Las células germinales transmiten estos genes indicadores, que de esta manera están presentes en todas las células del roedor recién nacido y se pueden utilizar para detectar la frecuencia de las mutaciones.

La **parte I** de este documento (capítulos 2-6) ofrece un breve panorama de las pruebas de genotoxicidad *in vivo*. Se explican los métodos utilizados en el diseño de los animales transgénicos, dando detalles de la construcción de ADN y de los métodos utilizados para su inserción en los animales receptores. Como ejemplos se describen modelos transgénicos - en particular, el modelo *lacI*, disponible comercialmente como ratón Big Blue® y rata Big Blue®, y el modelo *lacZ*, disponible comercialmente como Muta™Mouse -, así como otros modelos elaborados en fechas más recientes, por ejemplo el *λcII*, el *gpt* delta, el plásmido *lacZ* y el ΦX174.

El diseño del estudio es esencial para su validez a la hora de determinar la mutagenicidad positiva/negativa de un compuesto de prueba. La elección del gen, la especie y el tejido mutagénico destinatario debe basarse en los conocimientos previos sobre los parámetros farmacológicos/toxicológicos del agente de prueba. Dado que la selección de la dosis, el programa de dosificación y el tiempo de muestreo después del tratamiento varían para la detección óptima de la frecuencia de las mutaciones en distintos tejidos y agentes, se ha recomendado un protocolo que permita una detección óptima de todos los mutágenos, con independencia de su potencia o el tejido destinatario. La obtención de un resultado negativo utilizando un protocolo bien elaborado se debe considerar válida.

La **parte II** (capítulos 7-10) contiene una exposición general de los datos publicados sobre las sustancias químicas sometidas a prueba utilizando los modelos *lacI* y *lacZ*, compara éstos con los datos disponibles obtenidos mediante sistemas tradicionales y examina los resultados. Estos modelos se eligieron porque son los dos únicos sistemas con suficientes datos disponibles para realizar comparaciones y análisis.

Los limitados datos disponibles parecen indicar que hay un acuerdo significativo con respecto a los resultados obtenidos con la valoración del Muta™Mouse y el ratón (rata) Big Blue®. Cualquier diferencia observada entre la valoración del Muta™Mouse y el ratón (rata) Big Blue® es probable que se deba más al hecho de haber utilizado diseños experimentales distintos que a diferencias en la sensibilidad de los propios genes indicadores transgénicos.

Se compararon los resultados de las valoraciones de las mutaciones transgénicas con los obtenidos en la valoración con micronúcleos de médula ósea de ratón para 44 sustancias. Aunque la mayoría de los resultados fueron con frecuencia semejantes, como muchas de las sustancias sometidas a prueba eran carcinógenos potentes las valoraciones fueron complementarias, es decir, cuando se utilizaban ambas valoraciones se registraba una mejora significativa en la detección de carcinógenos. Este resultado parece confirmar la ventaja teórica de utilizar dos valoraciones que detectan efectos genotóxicos finales diferentes. Es también una ventaja clara la posibilidad de detectar mutaciones de genes en tejidos múltiples gracias a las valoraciones en animales transgénicos.

Aunque la prueba de mutación somática *in vivo* en ratones es un sistema normalizado de prueba de la genotoxicidad según las directrices de la Organización de Cooperación y Desarrollo Económicos (OCDE), raramente se ha utilizado este sistema en los últimos decenios para la detección de dichas mutaciones. Los resultados de una comparación de ambos sistemas en este documento puso de manifiesto que la valoración con ratones transgénicos tenía varias ventajas sobre la prueba de mutación somática *in vivo* en ratones y es un sistema de prueba adecuado para sustituirla en la detección de mutaciones en genes, pero no en cromosomas, *in vivo*.

A pesar de las diferencias en las propiedades mutacionales de los distintos modelos de mutágenos, las respuestas de los *loci* exógenos (transgén *lacI, lacZ*) y los *loci* endógenos (*Dlb-1, Hprt*) fueron en general cualitativamente semejantes después de los tratamientos de corta duración. Varios estudios parecen indicar que en tales condiciones la frecuencia más baja de mutantes somáticos en los genes endógenos puede proporcionar una mayor sensibilidad. Sin embargo, la comparación de los transgenes y los genes endógenos es difícil, debido a las diferencias que hay entre los protocolos experimentales óptimos para los distintos tipos de genes; en los transgenes neutros, la sensibilidad para la detección de mutaciones es mayor con los tiempos de administración más prolongados que se recomiendan actualmente.

Lo limitado de los datos sobre la comparación de la síntesis no programada de ADN con *lacI* y *lacZ* parece indicar que las valoraciones en animales transgénicos muestran una capacidad de predicción superior en comparación con la prueba de la síntesis no programada, que mide el daño en el ADN. Los resultados de las valoraciones en animales transgénicos (*lacI* y *lacZ)* con más de 50 sustancias químicas estaban en consonancia con los datos obtenidos *in vitro* sobre mutación de los genes, aberración cromosómica y medidas directas o indirectas del daño producido por esas sustancias en el ADN. Una ventaja importante de la valoración de las mutaciones en ratones/ratas transgénicos en comparación con otras pruebas de mutagenicidad *in vivo* es que se pueden detectar los casos de mutación en cualquier órgano. Por consiguiente, se realizó un análisis para determinar si, mediante la valoración de las mutaciones transgénicas, se podían predecir los órganos destinatarios en los estudios de carcinogenicidad. En la mayoría de los casos se encontraron mutaciones en los órganos destinatarios en los estudios de carcinogenicidad. Para varios carcinógenos supuestamente genotóxicos, los órganos investigados en las valoraciones de la mutagenicidad transgénica, que no eran órganos destinatarios en los estudios de la carcinogenicidad, dieron un resultado positivo. Debido a que esto ha ocurrido para varios compuestos, es poco probable que se pueda explicar por una especificidad insuficiente con respecto a los órganos destinatarios para la carcinogenicidad. Más bien nos lleva a la conclusión de que la genotoxicidad se expresa en varios órganos del organismo y que los tumores no se forman en todos esos órganos debido a otros factores. Los carcinógenos con un mecanismo de acción supuestamente no genotóxico suelen dar resultados negativos en las valoraciones en animales transgénicos. Hay muy pocos datos disponibles sobre las sustancias que dieron resultados negativos en las valoraciones de la carcinogenicidad con ratones. Sin embargo, para este pequeño número de sustancias no carcinógenas, los resultados en los ratones transgénicos también fueron negativos. Los datos disponibles parecen indicar que la sensibilidad y la capacidad de predicción positiva de las valoraciones transgénicas para la carcinogenicidad son elevadas.

La **parte III** (capítulo 11) describe estudios en los cuales se han utilizado valoraciones de las mutaciones transgénicas (en particular, en los modelos *lacI* y *lacZ* utilizando el sistema con roedores *cII* y *gpt* delta) como instrumento de investigación mecanicista. Gracias a la facilidad de la determinación de la secuencia del gen *cII* para los espectros mutacionales, cada vez se utiliza más para los estudios de determinación de secuencias en los modelos del Muta™Mouse y el Big Blue® en lugar de *lacI* y *lacZ*. También se utiliza el modelo *gpt* delta por la facilidad de determinación de la secuencia, y en particular porque detecta deleciones mucho más grandes que todos los demás, a excepción de la valoración con el plásmido *lacZ*.

Se han estudiado las mutaciones espontáneas en casi todas las valoraciones de la mutagenicidad en animales transgénicos: *lacZ*, *lacI* y *cII*, plásmido *lacZ* y ratones *gpt* delta. En todos los sistemas, el tipo predominante de mutación espontánea es la transición G:C→A:T, que se produce casi siempre en los lugares 5'-CpG, indicando que la desaminación de la 5-metilcitosina es el mecanismo principal de la mutagénesis.

Se ha estudiado la frecuencia y el carácter de las mutaciones espontáneas. Los factores que afectan a la tasa potencial de mutación son el punto de integración del transgén, la edad, el tejido y la estirpe. Alrededor de la mitad de todas las mutaciones se producen durante el desarrollo (y la mitad de éstas en el útero). En varios estudios se han examinado la frecuencia y el carácter de las mutaciones espontáneas con respecto a la edad en tejidos múltiples y se ha comprobado que, con la excepción de los estudios con plásmidos en el ratón, para la misma edad y tipo de tejido el espectro de los tipos de mutaciones en los animales adultos fue semejante. No presentó variaciones con las diferencias de sexo o antecedentes genéticos de los ratones. La frecuencia de las mutaciones en la línea germinal masculina fue de manera sistemática la más baja, manteniéndose básicamente inalterada en la vejez.

Se ha observado que las valoraciones en animales transgénicos son instrumentos útiles en el examen de paradigmas fundamentales de la toxicología genética. En estudios recientes utilizando estos sistemas se han abordado las cuestiones de 1) la relación dosis-respuesta de los carcinógenos genotóxicos y 2) la relación entre la formación de aductos de ADN, la frecuencia de las mutaciones y el cáncer en los roedores. Otra aplicación importante de estas valoraciones en roedores transgénicos ha sido la de estudios fundamentales sobre el origen de las mutaciones y la función de distintos procesos biológicos en su prevención. Entre ellos ha habido estudios sobre los mecanismos de reparación del ADN, la carcinogenicidad, el envejecimiento y las condiciones genéticas heredadas que afectan a estos procesos.

Si bien para la evaluación de las mutaciones de los genes *in vivo* no se considera obligatoria la obtención de espectros de las mutaciones a partir de los datos de la secuencia del ADN en el caso de resultados claramente positivos o negativos, son útiles para la investigación de los factores relacionados con el mecanismo de la mutagénesis. La capacidad para determinar la secuencia de las mutaciones inducidas en los genes indicadores transgénicos proporciona al investigador una información importante con respecto a varios aspectos de la mutación. Se citan ejemplos de estudios que demuestran la manera en que se pueden utilizar las valoraciones en animales transgénicos y el posterior análisis del espectro para examinar distintos aspectos de la actividad de los agentes mutagénicos: por ejemplo, 1) la corrección clonal y la corrección para mutaciones *ex vivo*, 2) las lesiones premutagénicas, 3) las respuestas de tejidos específicos, 4) la evaluación de sustancias genotóxicas que no interaccionan con el ADN, 5) la determinación de los componentes activos de mezclas, 6) la determinación de metabolitos activos y 7) la investigación de los mecanismos de las mutaciones de deleción *in vivo*.

En la **parte IV** (capítulos 12-14) se evalúa la función y el valor añadido potencial de la valoración de las mutaciones transgénicas en la toxicología y la evaluación del riesgo. Hasta el momento, la industria no ha utilizado de manera sistemática las valoraciones de la mutagenicidad transgénica en la investigación toxicológica, en gran parte porque no se ha elaborado todavía una directriz sobre pruebas de la OCDE. Recientemente se ha recomendado un protocolo armonizado internacionalmente (Thybaud et al., 2003), que debería constituir la base para dicha directriz.

El Grupo de Trabajo del IPCS recomienda la elaboración de una directriz de este tipo. Su utilidad se basa en parte en el hecho de que las valoraciones en animales transgénicos permiten detectar mutaciones de los genes. Si se utilizara dicho protocolo, un resultado negativo se podría considerar como fidedigno.

El Grupo de Trabajo del IPCS recomienda también que se incluyan las valoraciones de la mutagenicidad transgénica en el Sistema cualitativo del IPCS para la evaluación de la mutagenicidad y en otras estrategias de prueba.

Para futuras investigaciones, el Grupo de Trabajo del IPCS recomienda la verificación de varias sustancias no carcinógenas bien conocidas de acuerdo con un protocolo válido (por ejemplo, Thybaud et al., 2003). Se deberían recomendar las valoraciones de la mutagenicidad transgénica como instrumentos útiles para los estudios de la relación mecanicista entre las mutaciones y la carcinogénesis y para los estudios de la mutagénesis en las líneas germinales.

www.ingramcontent.com/pod-product-compliance
Lightning Source LLC
Chambersburg PA
CBHW071325210326
41597CB00015B/1356

9789241572330